AF777121

Encyclopaedia of Mathematics

Encyclopaedia of Mathematics

(Set of 3 Vol.)

Prof. Rajiv Kumar Gupta

CYBER TECH PUBLICATIONS

4264/3, Ansari Road, Daryaganj New Delhi - 110002 (India)

Phone : 011-23244078, 011-43559448, Fax : 011-23280028

Email : cybertechpublications@gmail.com

CYBER TECH PUBLICATION
4264/3, Ansari Road, Daryaganj
New Delhi - 110002 (India)
Phone : 011-23244078, 011-43559448,
Fax : 011-23280028
E-mail : cybertechpublications@gmail.com
Website : www.cybertechpublications.com

Encyclopaedia of Mathematics

First Edition : 2012

Rs. 2950.00

© Reserved

[No part of this publication may be reproduced, stored in a retrieval system or transmitted, in any form or by any means, mechanical, photocopying, recording or otherwise, without prior written permission of the publisher].

Printed at :

New Green Print Solutions, New Delhi 110078

PREFACE

This book has been written in general properties of the subject matter. This change is that paper division in the examination is different Universities. One of salient course syllabi features of the present edition of the book is the inclusion of a large number of typical worked out problems which will elucidate various abstract principles and theories discussed in the text. It covers the complete syllabus of physics prescribed by Technical Universities. The treatment given is simple, lucid and comprehensive. The language used is simple.

I shall, therefore, consider, my labours amply rewarded if the book receives warm reception and appreciation from the teachers and the taught. Though nothing can be claimed as original but the subject matter has been arranged in my own style. I hope, help the students to prepare for the final examination. I have spared no pains in applying my long experience of degree class teaching to make the book useful to the students.

—*Authors*

Contents

1

GEOMETRICAL MEANING OF THE DE MOIVRE'S THEOREM

INTRODUCTION

We know if $x + iy = r(\cos\theta + i\sin\theta)$, then it means (See Geometrical meaning of $x + iy$ in the chapter on complex numbers) a line of length r making an angle θ with OX or x-axis.

With the above meaning we can say that

$r[\cos(\theta + \phi) + i\sin(\theta + \phi)]$, means a line of the same length r making an angle $\theta + \phi$ with x-axis.

$\{r(\cos\theta + i\sin\theta)\}(\cos\phi + i\sin\phi) = r\{\cos(\theta + \phi) + i\sin(\theta + \phi)\}$

Hence, the effect of multiplying $r(\cos\theta + i\sin\theta)$ by $\cos\phi + i\sin\phi$ is to turn the line represented by $r(\cos\phi + i\sin\theta)$ or $x + iy$ through an angle f. Hence the quantity,

$(\cos\alpha + i\sin\alpha)(\cos\beta + i\sin\beta)(\cos\gamma + i\sin\gamma)(\cos\delta + i\sin\delta)$ means that the line represented by $(\cos\delta + i\sin\delta)$ is first turned through an angle γ, then through β and finally through a *i.e.*, altogether turned through an angle $\alpha + \beta + \gamma$ and this total operation as interpreted above gives the same line as

$$\{\cos(\alpha + \beta + \gamma) + i\sin(\alpha + \beta + \gamma)\}(\cos\delta + i\sin\delta)$$

A similar result can be derived for any number of factors.

Hence, by the De Moivre's Theorem the geometrical fact, that to turn a line through a number of angles successively has the same effect as turning through an angle equal to the sum of these angles, is expressed algebraically.

DIFINE OF THE DE MOIVRE'S THEOREM

Whatever be the value of n, positive or negative, integral or fractional, $\cos n\theta + i\sin n\theta$ is the value or one of the values of $(\cos\theta + \sin\theta)^n$.

Proof :

Case 1: When n is a *positive integer.*

By actual multiplication we have

$$(\cos \alpha + i \sin \alpha) (\cos \beta + \sin \beta)$$
$$= (\cos \alpha \cos \beta = \sin \alpha \sin \beta) + i (\cos \alpha \sin \beta = \sin \alpha \cos \beta)$$
$$= \cos (\alpha + \beta) + i \sin (\alpha + \beta)$$

Similarly, $(\cos \alpha + i \sin \alpha) (\cos \beta + i \sin \beta) (\cos \gamma + i \sin \gamma)$

$$= [\cos (\alpha + \beta) + i \sin (\alpha + \beta)] (\cos \gamma + i \sin \gamma)$$
$$= [\cos (\alpha + \beta) \cos \gamma - \sin (\alpha + \beta) \sin \gamma] + i [\sin (\alpha + \beta) \cos \gamma + \cos (\alpha + \beta) \sin \gamma]$$
$$= \cos (\alpha + \beta + \gamma) + i (\alpha + \beta + \gamma)$$

In this way we can prove that

$$(\cos \alpha + i \sin \alpha) (\cos \beta + i \sin \beta) (\cos \gamma + i \sin \gamma) (\cos \delta + i \sin \delta) \text{ ... to n factors}$$
$$= \cos (\alpha + \beta + \gamma + \delta + \text{... to n terms}) + i \sin (\alpha + \beta + \gamma + \delta + \text{... to n terms})$$

Putting $\alpha = \beta = \gamma = \delta = \ldots = \theta$, we have

$(\cos \theta + i \sin \theta) (\cos \theta + i \sin \theta)$ to n factors $= \cos n\theta + i \sin n\theta$

$$\Rightarrow \quad (\cos \theta + i \sin \theta)^n = \cos n\theta + i \sin n\theta$$

Case 2: When n is a negative integers, – m say). Where m itself is a positive integer.

In this case $(\cos \theta + i \sin \theta)^n = (\cos \theta + i \sin \theta)^{-m}$

$$= \frac{1}{(\cos \theta + i \sin \theta)^m}$$

$$= \frac{1}{\cos m\theta + i \sin m\theta} \quad \text{... by case I}$$

$$= \frac{1}{(\cos m\theta + i \sin m\theta)} \times \frac{(\cos m\theta - i \sin m\theta)}{(\cos m\theta + i \sin m\theta)},$$

multiplying both num. and denom. by $(\cos m\theta - i \sin m\theta)$

$$= \frac{\cos m\theta - i \sin m\theta}{\cos^2 m\theta - i^2 \sin^2 m\theta}$$

$$= \cos m\theta - i \sin m\theta \ \ldots \qquad \because i^2 = -1$$
$$= \cos (-m\theta) + i \sin (-m\theta)$$

$= \cos n\theta + i \sin n\theta$... $\because n = -m$

Case 3: When n is a fraction; say $n = \frac{p}{q}$, where θ is a positive integer and p is any integer positive or negative.

We have proved in cases I and II that

$(\cos\theta + i\sin\theta)^n = \cos n\theta + i\sin n\theta$, where n is any integer.

From here we can conclude that $\cos\theta + i\sin\theta$ is one of the values of $(\cos n\theta + i\sin n\theta)^{1/n}$, where n is any integer.

Now in this case 3.

$(\cos\theta + i\sin\theta)^n = (\cos\theta + i\sin\theta)^{p/q} = (\cos p\theta + i\sin p\theta)^{1/q}$ and as we have just shown above one of the values of

$$(\cos p\theta + i\sin p\theta)^{1/q} \text{ is } \cos\frac{p\theta}{q} + i\sin\frac{p\theta}{q}$$

Hence, De Moivre's Theorem is completly established

Note: $(\sin\theta + i\cos\theta)^n$ is not equal to $\sin n\theta + i\cos n\theta$.

SOLVED PROBLEMS

Example 1:

If $x = \cos\theta + i\sin\theta$; $y = \cos\phi + i\sin\phi$, *show that*

$$\frac{x-y}{x+y} = i\tan\frac{\theta-\phi}{2}.$$

Solution :

$$\text{L.H.S.} = \frac{x-y}{x+y} = \frac{(\cos\theta + i\sin\theta) - (\cos\phi + i\sin\phi)}{(\cos\theta + i\sin\theta) + (\cos\phi + i\sin\phi)}$$

$$= \frac{(\cos\theta - i\sin\phi) - i(\sin\theta - \sin\phi)}{(\cos\theta + i\sin\phi) + i(\sin\theta + \sin\phi)}$$

$$= \frac{\left[2\sin\frac{1}{2}(\theta+\phi)\sin\frac{1}{2}(\phi-\theta)\right] + i\left[2\cos\frac{1}{2}(\theta+\phi)\sin\frac{1}{2}(\theta-\phi)\right]}{\left[2\cos\frac{1}{2}(\theta+\phi)\cos\frac{1}{2}(\theta-\phi)\right] + i\left(2\sin\frac{1}{2}(\theta+\phi)\cos\frac{1}{2}(\theta-\phi)\right)}$$

$$= \frac{2\sin\frac{1}{2}(\theta-\phi)\left[\sin\frac{1}{2}(\theta+\phi) + i\cos\frac{1}{2}(\theta+\phi)\right]}{2\cos\frac{1}{2}(\theta-\phi)\left[\cos\frac{1}{2}(\theta+\phi) + i\sin\frac{1}{2}(\theta+\phi)\right]},$$

$$\because \sin\frac{1}{2}(\phi-\theta) = -\sin\frac{1}{2}(\theta-\phi)$$

$$= \frac{\sin\frac{1}{2}(\theta-\phi)\left[i\left\{i\sin\frac{1}{2}(\theta+\phi)+i\cos\frac{1}{2}(\theta+\phi)\right\}\right]}{\cos\frac{1}{2}(\theta-\phi)\left[\cos\frac{1}{2}(\theta+\phi)+i\sin\frac{1}{2}(\theta+\phi)\right]}, \quad \because i^2 = -1$$

$= \tan 1/2(\theta - \phi)\ [i] = i \tan\frac{1}{2}(\theta - \phi) = \text{R.H.S.}$ **Ans.**

Example 2:

If a denotes $\cos 2\alpha + i \sin 2a$, with similar expressions for b, c, d prove that

(i) $ab + cd = 2\cos(\alpha+\beta+\gamma-\delta)\,[\cos(\alpha+\beta+\gamma+\delta) + i\sin(\alpha+\beta+\gamma+\delta)]$

(ii) $(a+b)(c+d) = 4\cos(\alpha-\beta)\cos(\gamma-\delta\,[\cos(\alpha+\beta+\gamma+\delta) + i\sin(\alpha+\beta+\gamma+\delta)]$

Solution :

Let $b = \cos 2\beta + i\sin 2\beta$, $c = \cos 2\gamma + i\sin 2\gamma$

and $d = \cos 2\delta + i\sin 2\delta$

(i) $ab = (\cos 2\alpha + i\sin 2\alpha)(\cos 2\beta + i\sin 2\beta)$

$= (\cos 2\alpha\cos 2\beta - \sin 2\alpha\sin 2\beta) + i(\sin 2\alpha\cos 2\beta + \cos 2\alpha\sin 2\beta)$

$= \cos(2\alpha + 2\beta) + i\sin(2\alpha + 2\beta)$

Similarly it can be proved that

$cd = \cos(2\gamma + 2\delta) + i\sin(2\gamma + 2\delta)$

$\therefore ab + cd = [\cos(2\alpha+2\beta) + i\sin(2\alpha+2\beta)] + [\cos(2\gamma+2\delta) + i\sin(2\gamma+2\delta)]$

$= [\cos(2\alpha+2\beta) + \cos(2\gamma+2\delta] + i[\sin(2\alpha+2\beta) + \sin(2\gamma+2\delta)]$

$= [2\cos(\alpha+\beta+\gamma+\delta)\cos(\alpha+\beta-\gamma-\delta)] + i[2\sin(\alpha+\beta+\gamma+\delta)\cos(\alpha+\beta-\gamma-\delta)]$

$= 2\cos(\alpha+\beta-\gamma-\delta)\,[\cos(\alpha+\beta+\gamma+\delta) + i\sin(\alpha+\beta+\gamma+\delta)]$

Hence proved.

(ii) $a + b = (\cos 2\alpha + i \sin 2\alpha) + (\cos 2\beta + i \sin 2\beta)$

$= (\cos 2\alpha + \cos 2\beta) + i (\sin 2\alpha + \sin 2\beta)$

$= 2 \cos (\alpha - \beta) \cos (\alpha + \beta) + i [2 \sin (\alpha + \beta) \cos (\alpha - \beta)]$

$= 2 \cos (\alpha - \beta) [\cos (\alpha + \beta) + i \sin (\alpha + \beta)]$.

Similarly we can prove that

$c + d = 2 \cos (\gamma - \delta) [\cos (\gamma + \delta) + i \sin (\gamma + \delta)]$

$\therefore (a + b) (c + d) = [2 \cos (\alpha - \beta) \{\cos (\alpha + \beta) + i \sin (\alpha + \beta)\}]$
$\times [2 \cos (\gamma - \delta) \{\cos (\gamma + \delta) + i \sin (\gamma + \delta)\}]$

$= 4 \cos (\alpha - \beta) \cos (\gamma + \delta) [\cos (\alpha + \beta) + i \sin (\alpha + \beta)]$
$\times [\cos (\gamma + \delta) + i \sin (\gamma + \delta)]$

$= \cos (\alpha - \beta) \cos (\gamma - \delta) [\{\cos (\alpha + \beta) \cos (\gamma + \delta) - \sin (\alpha + \beta)$
$\times \sin (\gamma + \delta)\} + i \{\sin (\alpha + \beta) \cos (\gamma + \delta) + \cos (\alpha + \beta) \sin (\gamma + \delta)\}]$

$= 4 \cos (\alpha - \beta) \cos (\gamma - \delta) [\cos (\alpha + \beta + \gamma + \delta) + i \sin (\alpha + \beta + \gamma + \delta)]$

Hence proved.

Example 3:

If $\cos \theta = a + \frac{1}{a}$: $2 \cos \phi = b + \frac{1}{b}$; *etc. show that*

(i) $abc... + \frac{1}{abc....} = 2 \cos (\theta + \phi + ...)$

(ii) $a^p b^q c^r + ... + \frac{1}{a^p b^q c^r.....}\ 2 \cos (p\theta + q\phi + ...)$

Solution :

Given $a + (1/a) = 2 \cos \theta$

$\Rightarrow \qquad a^2 + 1 = 2 \cos \theta$

$\Rightarrow \qquad a^2 - 2\alpha \cos \theta + 1 = 0$

$$\Rightarrow \qquad a = \frac{2 \cos \theta + \sqrt{(4 \cos^2 \theta - 4)}}{2}$$

$$= \cos \theta + \sqrt{(\cos^2 \theta - 1)} = \cos \theta + \sqrt{(-\sin^2 \theta)}.$$

$\Rightarrow \qquad a = \cos \theta + i \sin \theta. \qquad \because i^2 = -1$

Similarly $b = \cos \phi + i \sin \phi$, etc.

(i) $abc. = (\cos \theta + i \sin \theta) (\cos \phi + i \sin \phi) (\cos \phi + i \sin \phi) (\cos \psi + i \sin \psi)...$

$= \cos(\theta + \phi + \psi + ...) + i \sin(\theta + \phi + \psi + ...)$

and $\frac{1}{abc...} = (abc)...)^{-1}$

$= [\cos(\theta + \phi + \psi + ...) + i \sin(\theta + \phi + \psi + ...)]^{-1}$

$= \cos(\theta + \phi + \psi + ...) - i \sin(\theta + \phi + \psi + ...)$

$\therefore abc + ... \frac{1}{abc...} = [\cos(\theta + \phi + \psi + ...) + i \sin(\theta + \phi + \psi + ...)]$

$+ [\cos(\theta + \phi + \psi + ...) - i \sin(\theta + \phi + \psi + ...)]$

$= 2 \cos(\theta + \phi + \psi + ...)$

(ii) $a^p = (\cos\theta + i \sin\phi)^p = \cos p\theta + i \sin p\theta;$

$b^q = (\cos\phi + i \sin\phi)^q = \cos q\phi + i \sin q\phi;$ ... etc.

$\therefore a^p b^q c^r ... = (\cos p\theta + i \sin p\theta)(\cos q\phi + i \sin q\phi)...$

$= \cos(p\theta + q\phi + ...) + i \sin(p\theta + q\phi + ...)$

And $1/a^p b^q c^r ... = (a^p b^q c^r ...)^{-1}$

$= [\cos(p\theta + q\phi + ...) + i \sin(p\theta + q\phi ...)]^{-1}$

$= \cos(p\theta + q\phi + ...) - i \sin(p\theta + q\phi + ...)$

$\therefore a^p b^q c^r ... + \frac{1}{a^p b^q c^r ...} = [\cos(p\theta + q\phi + ...) + i \sin(p\theta + q\phi + ...)]$

$+ [\cos(p\theta + q\phi + ...) - i \sin(p\theta + q\phi + ...)]$

$= 2 \cos(p\theta + q\phi + ...)$ **Hence proved.**

Example 4:

Show that

$$(1+\cos\theta + i\sin\theta)^n + (1+\cos\theta - i\sin\theta)^n = 2^{n+1}\cos^n\left(\frac{1}{2}\theta\right).\cos\left(\frac{1}{2}n\theta\right).$$

Solution :

L.H.S. $= (1 + \cos\theta + i \sin\theta)^n + (1 + \cos\theta - i \sin\theta)^n$

$= [2\cos^2 \frac{1}{2}\theta + i . 2 \sin \frac{1}{2}\theta \cos \frac{1}{2}\theta]^n$

$+ [2\cos^2 \frac{1}{2}\theta - i . 2 \sin \frac{1}{2}\theta \cos \frac{1}{2}\theta]^n$

$= 2^n \cos^n\left(\frac{1}{2}\theta\right) [\{\cos \frac{1}{2}\theta + i \sin \frac{1}{2}\theta\}^n + \{\cos \frac{1}{2}\theta - \sin \frac{1}{2}\theta\}^n]$

$$= 2^n \cos^n\left(\frac{1}{2}\theta\right) \; [(\cos \frac{1}{2} n\theta + i \sin \frac{1}{2} n\theta) + (\cos \frac{1}{2} n\theta - \sin \frac{1}{2} n\theta)]$$

$$= 2^n \cos^n\left(\frac{1}{2}\theta\right) \; [2 \cos \frac{1}{2} n\theta] = 2^{n+1} \cos^n\left(\frac{1}{2}\theta\right) \cos\left(\frac{1}{2} n\theta\right) = \text{R.H.S.}$$

Example 5:

Show that

$$\left(\frac{1+\cos\phi+i\sin\phi}{1+\cos\phi-i\sin\phi}\right)^n = \cos n\phi + i \sin n\phi.$$

Solution :

$$\text{L.H.S.} = \left[\frac{2\cos^2\frac{1}{2}\phi + i\,2\sin\frac{1}{2}\phi\cos\frac{1}{2}\phi}{2\cos^2\frac{1}{2}\phi - i.2\sin\frac{1}{2}\phi\cos\frac{1}{2}\phi}\right]^n$$

$$= \left[\frac{\cos\frac{1}{2}\phi + i\sin\frac{1}{2}\phi}{\cos\frac{1}{2}\phi - i\sin\frac{1}{2}\phi}\right]^n$$

$$= (\cos \frac{1}{2}\phi + i \sin \frac{1}{2}\phi)^n (\cos \frac{1}{2}\phi - i \sin \frac{1}{2}\phi)^{-n}$$

$$= (\cos \frac{1}{2} n\phi + i \sin \frac{1}{2} n\phi) (\cos \frac{1}{2} n\phi + i \sin \frac{1}{2} n\phi)$$

$$= (\cos \frac{1}{2} n\phi + i \sin \frac{1}{2} n\phi)^2$$

$$= (\cos n\phi + i \sin n\phi) = \text{R.H.S.}$$

Example 6:

Show that

$$\left[\frac{1+\sin\phi+i\cos\phi}{1+\sin\phi-i\cos\phi}\right]^n = \cos(\frac{1}{2}n\pi - n\phi) + i\sin(\frac{1}{2}n\pi - n\phi).$$

Solution :

$$\text{L.H.S.} = \left[\frac{1+\sin\phi+i\cos\phi}{1+\sin\phi \;\; i\cos\phi}\right]^n$$

$$= \left[\frac{1+\cos\theta+i\sin\theta}{1+\cos\theta-i\sin\theta}\right]^n, \text{ putting } \phi = \frac{1}{2}\pi - \theta$$

$$= \left[\frac{2\cos^2\frac{1}{2}\theta + i.2\sin\frac{1}{2}\theta\cos\frac{1}{2}\theta}{2\cos^2\frac{1}{2}\theta - i.2\sin\frac{1}{2}\theta\cos\frac{1}{2}\theta}\right]^n$$

$$\because\ 1+\cos\theta = 2\cos^2\frac{1}{2}\theta \text{ and } \sin\theta = 2\sin\frac{1}{2}\theta\cos\frac{1}{2}\theta$$

$$= \left[\frac{2\cos\frac{1}{2}\theta\left(\cos\frac{1}{2}\theta + i\sin\frac{1}{2}\theta\right)}{2\cos\frac{1}{2}\theta\left(\cos\frac{1}{2}\theta - i\sin\frac{1}{2}\theta\right)}\right]^n = \left[\frac{\left(\cos\frac{1}{2}\theta + i\sin\frac{1}{2}\theta\right)}{\left(\cos\frac{1}{2}\theta - i\sin\frac{1}{2}\theta\right)}\right]^n$$

$$= [(\cos\frac{1}{2}\theta + i\sin\frac{1}{2}\theta) \times (\cos\frac{1}{2}\theta - i\sin\frac{1}{2}\theta)^{-1}]^n$$

$$= [(\cos\frac{1}{2}\theta + i\sin\frac{1}{2}\theta)(\cos\frac{1}{2}\theta + i\sin\frac{1}{2}\theta)]^n$$

$$= [(\cos\frac{1}{2}\theta + i\sin\frac{1}{2}\theta)^2]^n$$

$$= (\cos\theta + i\sin\theta)^n = \cos n\theta + i\sin n\theta$$

$$= \cos n\left(\frac{1}{2}\pi - \phi\right) + i\sin n\left(\frac{1}{2}\pi - \phi\right), \quad \because\ \theta = \frac{1}{2}\pi - \theta$$

$$= \cos\left(\frac{1}{2}n\pi - n\phi\right) + i\sin\left(\frac{1}{2}n\pi - n\phi\right) = \text{R.H..S.}$$

Example 7:

Show that 4n the power of $(1 + 7i)/(2 - i)^2$ is equal to $(-4)^n$, where n is a positive integer.

Solution :

$$\frac{1+7i}{(2-i)^2} = \frac{(1+7i)(2+i)^2}{(2-i)^2(2+i)^2}, \text{ multiplying numerator and denominator by}$$

$(2 + i)^2$

$$= \frac{(1+7i)\,(4+4i+i^2)}{(4-i^2)^2} = \frac{(1+7i)\,(3+4i)}{(4+1)^2} \qquad \because\ i^2 = -1. \qquad ...(i)$$

$$= \frac{3+4i+21i+28i^2}{25} = \frac{3+25i-28}{25} = -1 + i.$$

Let $-1 + i = r(\cos\theta + i\sin\theta)$. ...(ii)

Then $r\cos\theta = -1$, $r\sin\theta = 1$, equating real and imaginary parts on both sides

Squaring and adding these we get $r^2 = 2$ ...(iii)

Dividing we get $\tan\theta = -1$ which implies $\theta = -\frac{1}{4}\pi$, ...(iv)

taking the principal value of θ which lies between $-\pi$ and $+\pi$

$\therefore$ From (i) we get

$$\left[\frac{1+7i}{(2-i)^2}\right]^{4n} = [-1 + i]^{4n}, \text{ from (i)}$$

$= [r(\cos\theta + i\sin\theta)]^{4n}$, from (ii)

$= (r)^{4n}(\cos\theta + i\sin\theta)^{4n}$

$= (r^2)^{2n}(\cos 4n\theta + i\sin 4n\theta)$ **(Note)**

$= (2)^{2n}\{\cos(-n\pi + i\sin n\pi\}$, where n is a positive integer.

$= 4^n\{(-1)^n + i(0)\}$, $\because \cos n\pi = (-1)^n$, $\sin n\pi = 0$.

$= 4^n(-1)^n = (-4)^n$. **Hence proved.**

Example 8:

Prove that

$(a + ib)^{m\,n} + (a - ib)^{m\,n} = 2(a^2 + b^2)^{m\,2n}\cos[(m/n)\tan^{-1}(b/a)]$.

Solution :

Let $a = r\cos\theta$ and $b = r\sin\theta$

then squaring and adding we have $r^2 = a^2 + b^2 \Rightarrow r = \sqrt{(a^2 + b^2)}$ ...(i)

and dividing we get $\tan\theta = b/a \Rightarrow \theta = \tan^{-1}(b/a)$ ...(ii)

$\therefore$ L.H.S. $= (a + ib)^{m/n} + (a - ib)^{m/n}$

$= [r\cos\theta + ir\sin\theta]^{m/n} + [r\cos\theta - ir\sin\theta]^{m/n}$,

putting $a = r\cos\theta$ and $b = r\sin\theta$

$$= [r(\cos\theta + i\sin\theta)^{m/n} + [r(\cos\theta - i\sin\theta)]^{m/n}$$

$$= r^{m/n}(\cos\theta + i\sin\theta)^{m/n} + r^{m/n}(\cos\theta - i\sin\theta)^{m/n}$$

$$= r^{m/n}\left[\left(\cos\frac{m\theta}{n} + i\sin\frac{m\theta}{n}\right) + \left(\cos\frac{m\theta}{n} - i\sin\frac{m\theta}{n}\right)\right]$$

$$= r^{m/n}\left[2\cos\frac{m\theta}{n} = 2r^{m/n}\cos\frac{m\theta}{n}\right]$$

$$= 2\left[\sqrt{(a^2+b^2)}\right]^{m/n}\cos\frac{m}{n}\left(\tan^{-1}\frac{b}{a}\right), \text{ from (i) and (ii)}$$

$$= 2(a^2+b^2)^{m/2n}\cos\frac{m}{n}\left(\tan^{-1}\frac{b}{a}\right) = \text{R.H.S.}$$

Example 9:

Show that $(1+i)^n + (1-i)^n = 2^{(n/2)+1}\cos\left(\frac{1}{4}n\pi\right)$.

Solution :

Let $1 + i = r(\cos\theta + i\sin\theta)$

Equating real and imaginary parts, we have

$r\cos\theta = 1$ and $r\sin\theta = 1$ ...(i)

Squaring and adding we have $r^2 = 1 + 1 = 2 \Rightarrow r = \sqrt{2}$

$\therefore$ From (i), $\cos\theta = 1/\sqrt{2}$ and $\sin\theta = 1/\sqrt{2}$, both of the equations are satisfied by $\theta = \frac{1}{4}\pi$

Hence $1 + i = \sqrt{2}\left(\cos\frac{1}{4}\pi + i\sin\frac{1}{4}\pi\right)$

Similarly $1 - i = \sqrt{2}\left(\cos\frac{1}{4}\pi - i\sin\frac{1}{4}\pi\right)$

Substituting these values of $(1+i)$ and $(1-i)$, we have

$$\text{L.H.S.} = \left[\sqrt{2}\left(\cos\frac{1}{4}\pi + i\sin\frac{1}{4}\pi\right)\right]^n + \left[\sqrt{2}\left(\cos\frac{1}{4}\pi - i\sin\frac{1}{4}\pi\right)\right]^n$$

$$= \left[2^{n/2}\left(\cos\frac{1}{4}n\pi + i\sin\frac{1}{4}n\pi\right)\right] - \left[2^{n/2}\left(\cos\frac{1}{4}n\pi - i\sin\frac{1}{4}n\pi\right)\right]$$

$$= 2^{n/2}\left[\left(\cos\frac{1}{4}n\pi + i\sin\frac{1}{4}n\pi\right) + \left(\cos\frac{1}{4}n\pi - i\sin\frac{1}{4}n\pi\right)\right]$$

$$= 2^{n/2}\left[2\cos\frac{1}{4}n\pi\right] = 2^{(n/2)+1}\cos\frac{1}{4}n\pi = \text{R.H.S.}$$

Example 10:

If $x_r = \cos(\pi/2^r) + i\sin(\pi/2^r)$, prove that $x_1x_2x_3$...ad. inf. = – 1.

Solution :

Given that $x_r = \cos(\pi/2^r) + i\sin(\pi/2^r)$

Putting r = 1, 2, 3, we have

$$x_1 = \cos\left(\frac{1}{2}\pi\right) + i\sin\left(\frac{1}{2}\pi\right),$$

$$x_2 = \cos\left(\frac{\pi}{2^2}\right) + i\sin\left(\frac{\pi}{2^2}\right),$$

$$x_3 = \cos\left(\frac{\pi}{2^3}\right) + i\sin\left(\frac{\pi}{2^3}\right), \text{...etc.}$$

$$x_2x_3x_3.. = \left[\cos\left(\frac{1}{2}\pi\right) + i\sin\left(\frac{1}{2}\pi\right)\right] \times$$

$$\left[\cos\left(\frac{\pi}{2^2}\right) + i\sin\left(\frac{\pi}{2^3}\right)\right] \times \left[\cos\left(\frac{\pi}{2^2}\right) + i\sin\left(\frac{\pi}{2^3}\right)\right] \times \text{...ad. inf.}$$

$$= \cos\left(\frac{1}{2}\pi + \frac{\pi}{2^2} + \frac{\pi}{2^3} + \text{...ad.inf.}\right) + i\sin\left(\frac{1}{2}\pi + \frac{\pi}{2^2} + \frac{\pi}{2^3} + \text{...ad.inf.}\right)$$

$$= \cos\left[\frac{1}{2}\pi\left(1 + \frac{1}{2} + \frac{1}{2^2} + \text{...ad.inf.}\right)\right] + i\sin\left[\frac{1}{2}\pi\left(1 + \frac{1}{2} + \frac{1}{2^2} + \text{...ad.inf.}\right)\right]$$

$$= \cos\left[\frac{\frac{1}{2}\pi}{1-\frac{1}{2}}\right] + i\sin\left[\frac{\frac{1}{2}\pi}{1-\frac{1}{2}}\right], \because 1 + r + r^2 + \text{... ad. inf.} = \frac{\text{"a"}}{1-r}$$

$= \cos\pi + i\sin\pi = -1 - i(0) = -1.$ **Hence proved.**

Example 11:

If $(a_1 + ib_1)(a_2 + ib_2)(a_3 + ib_3) \ldots (a_n + ib_n) = A + iB$, *prove that*

(i) $\left(a_1^2+b_1^2\right)\left(a_2^2+b_2^2\right)\left(a_3^2+b_3^2\right)\ldots\left(a_n^2+b_n^2\right) = A^2 + B^2$ *and*

(ii) $\tan^{-1}\frac{b_1}{a_1} + \tan^{-1}\frac{b_2}{a_2} + \tan^{-1}\frac{b_3}{a_3} + \ldots + \tan^{-1}\frac{b_n}{a_n} = \tan^{-1}\frac{B}{A}$

Solution :

Let $a_1 + ib_1 = r_1(\cos\theta_1 + i\sin\theta_1)$;

$$a_2 + ib_2 + r_2(\cos\theta_2 + i\sin\theta_2),$$

and similar other expressions.

Then $r_1 = \sqrt{\left(a_1^2+b_1^2\right)}$, $r_2 = \sqrt{\left(a_2^2+b_2^2\right)}$ etc.

and $\theta_1 = \tan^{-1}\frac{b_1}{a_1}$, $\theta_2 = \tan^{-1}\frac{b_2}{a_2}$ etc. ...(a)

Now it is given that

$$(a_1 + ib_1)(a_2 + ib_2)(a_3 + ib_3) \ldots (a_n + ib_n) = A + iB$$

$\Rightarrow [r_1(\cos\theta_1 + i\sin\theta_1)][r_2(\cos\theta_2 + i\sin\theta_2)]$

$\ldots [r_n(\cos\theta_n + i\sin\theta_n)] = A + iB,$

substituting the values of $a_1 + ib_1$ etc.

$\Rightarrow r_1r_2r_3\ldots r_n[(\cos\theta_1 + i\sin\theta_1)(\cos\theta_2 + i\sin\theta_2)\ldots$

$\ldots(\cos\theta_n + i\sin\theta_n)] = A + iB$

$\Rightarrow r_1r_2r_3\, r_n[\cos(\theta_1 + \theta_2 + \ldots + \theta_n) + i\sin(\theta_1 + \theta_2 + \ldots + \theta_n)] = A +$
iB

Equating real and imaginary parts on both sides, we get

$A = r_1r_2r_3\ldots r_n \cos(\theta_1 + \theta_2 + \theta_3 + \ldots + \theta_n)$ ...(b)

$B = r_1r_2r_3\ldots r_n \sin(\theta_1 + \theta_2 + \theta_3 + \ldots + \theta_n)$ ...(c)

(i) Squaring and adding (b) and (c) we get

$$A^2 + B^2 = r_1^2r_2^2r_3^3\ldots r_n^2 = \left(a_1^2+b_1^2\right)\left(a_2^2+b_2^2\right)\ldots\left(a_n^2+b_n^2\right) \text{ from (a)}$$

(ii) Dividing (c) by (b) we get $B/A = \tan(\theta_1 + \theta_2 + \ldots + \theta_n)$

$\Rightarrow \tan^{-1}(B/A) = \theta_1 + \theta_2 + \ldots\theta_n$

$= \tan^{-1}\frac{b_1}{a_1} + \tan^{-1}\frac{b_2}{a_2} + \tan^{-1}\frac{b_3}{a_3} + \ldots,$ from (a)

Example 12:

If $\left(1+i\frac{x}{a}\right)\left(1+i\frac{x}{b}\right)\left(1+i\frac{x}{c}\right)\ldots A + iB$ *then prove that*

(i) $\left(1+\frac{x^2}{a^2}\right)\left(1+\frac{x^2}{b^2}\right)\left(1+\frac{x^2}{c^2}\right)\ldots A^2 + B^2$

(ii) $tan^{-1}\frac{x}{a}+tan^{-1}\frac{x}{b}+tan^{-1}\frac{x}{c}+\ldots=tan^{-1}\frac{B}{A}$.

Solution :

Let $\frac{x}{a} = \tan\alpha,\ \frac{x}{b} = \tan\beta,\ \frac{x}{c} = \tan\gamma$, ...etc. ...(a)

Then the give result becomes

$$(1 + i\tan\alpha)(1 + i\tan\beta)(1 + i\tan\gamma)\ldots = A + iB$$

$$\Rightarrow \left(1+i\frac{\sin\alpha}{\cos\alpha}\right)\left(1+i\frac{\sin\beta}{\cos\beta}\right)\left(1+i\frac{\sin\gamma}{\cos\gamma}\right)\ldots = A + iB$$

$$\Rightarrow \left(\frac{\cos\alpha + i\sin\alpha}{\cos\alpha}\right)\left(\frac{\cos\beta + i\sin\beta}{\cos\beta}\right)\left(\frac{\cos\gamma + i\sin\gamma}{\cos\gamma}\right)\ldots = A + iB$$

$$\Rightarrow \frac{1}{\cos\alpha\cos\beta\cos\gamma\ldots}(\cos\alpha + i\sin\alpha)(\cos\beta + i\sin\beta)$$

$\times (\cos\gamma + i\sin\gamma)\ A + iB$

$\times \sec\alpha\sec\beta\sec\gamma\ldots[\cos(\alpha + \beta + \gamma + \ldots) + i\sin(\alpha + \beta + \gamma + \ldots)]$

$= A + iB$

Equating real and imaginary parts on both sides, we get

$$A = \sec\alpha\sec\beta\sec\gamma\ldots\cos(\alpha + \beta + \gamma + \ldots) \quad \ldots(b)$$

$$B = \sec\alpha\sec\beta\sec\gamma\ldots\sin(\alpha + \beta + \gamma + \ldots) \quad \ldots(c)$$

(i) Squaring and adding (b) and (c) we get

$A^2 + B^2 = \sec^2\alpha\ \sec^2\beta,\ \sec^2\gamma\ldots,$

$= (1 + \tan^2\alpha)(1 + \tan^2\beta)(1 + \tan^2\beta)(1 + \tan^2\gamma)\ldots$

$= \left(1+\frac{x^2}{a^2}\right)\left(1+\frac{x^2}{b^2}\right)\left(1+\frac{x^2}{c^2}\right)\ldots$, from (a)

(ii) Dividing (c) by (b) we get $(B/A) = \tan(\alpha + \beta + \gamma + \ldots)$

$\Rightarrow \quad \tan^{-1}(B/A) = \alpha + \beta + \gamma + \ldots$

$$= \tan^{-1}\frac{x}{a} + \tan^{-1}\frac{x}{b} + \tan^{-1}\frac{x}{c} + \ldots, \text{ from (a)}$$

Hence proved.

Example 13:

$\cos\alpha + \cos\beta + \cos\gamma = \sin\alpha + \sin\beta + \sin\gamma = 0$, prove that
$\cos 3\alpha + \cos 3\beta + \cos 3\gamma = 3\cos(\alpha + \beta + \gamma)$ and
$\sin 3\alpha + \sin 3\beta + \sin 3\gamma = 3\sin(\alpha + \beta + \gamma)$.

Solution :

We know that if $a + b + c = 0$, $a^3 + b^3 + c^3 - 3abc = 0$

$\Rightarrow \quad a^3 + b^3 + c^3 = 3abc$ ...(i)

Now let $\alpha \cos\alpha + i\sin\alpha : b = \cos\beta + i\sin\beta$ and

$c = \cos\gamma + i\sin\gamma$

Then $a + b + c = (\cos\alpha + i\sin\alpha) + (\cos\beta + i\sin\beta) + (\cos\gamma + i\sin\gamma)$.

$= (\cos\alpha + \cos\beta + \cos\gamma) + i(\sin\alpha + \sin\beta + \sin\gamma)$

$\Rightarrow a + b + c = (0) + i(0)$ ...(given)

$= 0$

$\therefore$ From (i) we have $a^3 + b^3 + c^3 = 3abc$

$\Rightarrow (\cos\alpha + i\sin\alpha)^3 + (\cos\beta + i\sin\beta)^3 + (\cos\gamma + i\sin\gamma)^3$

$= 3(\cos\alpha + i\sin\alpha)(\cos\beta + i\sin\beta)(\cos\gamma + i\sin\gamma)$

$\Rightarrow (\cos 3\alpha + i\sin 3\alpha) + (\cos 3\beta + i\sin 3\beta) + (\cos 3\gamma + i\sin 3\gamma)$

$= 3[\cos(\alpha + \beta + \gamma) + i\sin(\alpha + \beta + \gamma)]$

$\Rightarrow (\cos 3\alpha + \cos 3\beta + \cos 3\gamma) + i(\sin 3\alpha + \sin 3\beta + \sin 3\gamma)$

$= 3[\cos(\alpha + \beta + \gamma) + i\sin(\alpha + \beta + \gamma)]$

Equating real and imaginary parts on both sides we get

$\cos 3\alpha + \cos 3\beta + \cos 3\gamma = 3\cos(\alpha + \beta + \gamma)$

and $\sin 3\alpha + \sin 3\beta + \sin 3\gamma = 3\sin(\alpha + \beta + \gamma)$. **Hence proved.**

Example 14:

Find the general value of q with satisfies the equations
$(\cos\theta + i\sin q)(\cos 2\theta + i\sin 2\theta)\ldots(\cos r\theta + i\sin r\theta) = 1$

Solution :

$(\cos\theta + i\sin\theta)(\cos 2\theta + i\sin 2\theta)(\cos 3\theta + i\sin 3\theta)\ldots$

$\times (\cos r\theta + i\sin r\theta) = 1$

$\Rightarrow$ $\cos(\theta + r\theta + 3\theta + ... + r\theta) + i \sin(\theta + 2\theta + ... + r\theta) = 1$

$\Rightarrow$ $\cos(1 + 2 + 3 + ... + r)\,\theta + i \sin(1 + 2 + 3 + ... + r)\,\theta = 1$

$$\Rightarrow \cos\left[\frac{1}{2}r\,(r+1)\,\theta\right] + i \sin\left[\frac{1}{2}r\,(r+1)\,\theta\right] = 1$$

$$\because\ 1 + 2 + 3 + ... + r = \frac{1}{2}r\,(r + 1)$$

Equating real parts on both sides, we get

$$\cos\left[\frac{1}{2}r\,(r+1)\,\theta\right] = 1 = \cos 0,$$

$\therefore$ $\frac{1}{2}r\,(r + 1)\,\theta = 2n\pi \pm 0$, when n is zero or any integer,

$$\Rightarrow \quad r\,(r + 1)\,\theta = 4n\pi$$

$\Rightarrow$ $\theta = \dfrac{4n\pi}{r\,(r+1)}$, where n is zero or any integer. **Ans.**

Example 15:

Find the general value of θ which satisfies the equation

$$(\cos\theta + i \sin\theta)(\cos 3\theta + i \sin 3\theta)...\{\cos(2r - 1)\,\theta + i \sin(2r - 1)\,\theta\} = 1.$$

Solution :

$(\cos\theta + i \sin\theta)(\cos 3\theta + i \sin 3\theta)...$

$... \{\cos(2r - 1)\,\theta + i \sin(2r - 1\theta) = 1$

$\Rightarrow$ $\cos\{\theta + 3\theta + 5\theta + ... + (2r - 1)\,\theta\} + i \sin\{\theta + 3\theta + 5\theta + ... + (2r - 1)\,\theta\} = i$

$\Rightarrow$ $\cos\{1 + 3 + 5 + ... + (2r - 1)\}\,\theta + i \sin\{1 + 3 + 5 + ... + (2r - 1)\}\,\theta = 1$

$$\Rightarrow \cos\left[\frac{1}{2}r\,\{2 + (r-1)\,2\}\right]\theta + i \sin\left[\frac{1}{2}r\,\{2 + (r-1)\,2\}\right]\theta = 1$$

$\Rightarrow$ $\cos(r^2\theta) + i \sin(r^2\theta) = 1$

Equating real parts on both sides, we get

$\cos(r^2\theta) = 1 = \cos 0$

$\therefore$ $r^2\theta\ 2n\pi \pm 0$, where n is zero or any integer.

$\Rightarrow$ $\theta = 2n\pi/r^2$, where n is zero or any integer. **Ans.**

Example 16:

Find the equations whose roots are the nth powers of the roots of the equation $x^2 - 2x \cos \theta + 1 = 0$.

Solution :

The equation is $x^2 - 2x \cos \theta + 1 = 0$

$$\therefore \quad x = \frac{2\cos\theta + \sqrt{(4\cos^2\theta - 4)}}{2} = \cos\theta \pm \sqrt{(\cos^2\theta - 1)}$$

$$= \cos\theta \pm \sqrt{(-\sin^2\theta)} = \cos\theta \pm i\sin\theta$$

$\therefore$ The roots of the given equation are $\cos\theta + i\sin\theta$ and $\cos\theta - i\sin\theta$. Hence the roots of the required equation are

$$(\cos\theta + i\sin\theta)^n \text{ and } (\cos\theta - i\sin\theta)^n$$

i.e., $(\cos n\theta + i\sin n\theta)$ and $(\cos n\theta - i\sin n\theta)$

$\therefore$ The sum of the roots of the required equation

$$= (\cos n\theta + i\sin n\theta) + (\cos n\theta - i\sin n\theta) = 2\cos n\theta$$

and the product of the required equation

$$= (\cos n\theta - i\sin n\theta) \times (\cos n\theta - i\sin n\theta)$$

$$= \cos^2 n\theta - i2.\ \sin2\ n\theta$$

$$= \cos^2 n\theta + \sin^2 n\theta \quad \dots \quad \because i^2 = -1$$

$$= 1$$

$\therefore$ The required equation is

$x^2 - x$ (sum of the roots) + (product of the roots) = 0

$\Rightarrow x^2 - 2x\cos n\theta + 1 = 0.$ **Ans.**

Example 17:

If $x = \cos\theta + i\sin\theta$ *and* $\sqrt{1 - c^2} = nc - 1$, *show that* $1 + c\cos\theta = \frac{c}{2n}(1+nx)\left(1+\frac{n}{x}\right)$.

Solution :

If $x = \cos\theta + i\sin\theta$,

then $\frac{1}{x} = (\cos\theta + i\sin\theta)^{-1} = \cos\theta - i\sin\theta$ ...(i)

Also $\sqrt{(1-c^2)} = nc - 1$ or $1 - c^2 = (nc - 1)^2$, squaring both sides

$\Rightarrow \quad 1 - c^2 = n^2c^2 - 2nc + 1$

$\Rightarrow \quad c^2 (n^2 + 1) = 2nc$

$\Rightarrow \quad c (n^2 + 1) = 2n$ ...(ii)

Now R.H.S. $= \frac{c}{2n}(1+nx)\left(1+\frac{n}{x}\right)$.

$= \frac{c}{2n}\left(1+nx+\frac{n}{x}+n^2\right) = \frac{c}{2n}(1+n^2) + \frac{c}{2n}n\left(x+\frac{1}{x}\right)$

$= \frac{c(1+n^2)}{2n} + \frac{c}{2}$ [(cos θ + i sin θ) + (cos θ − i sin θ)], from (1)

$= \frac{2n}{2n} + \frac{c}{2}$ [2 cos θ]... $\because$ c (1 + n²) = 2n, from (ii)

$= 1 + c\cos\theta$ = L.H.S. **Hence proved.**

Example 18:

Find the value of x such that

$$\frac{(x+\alpha)^n-(x+\beta)^n}{\alpha-\beta} = \frac{\sin n\theta}{\sin^n \theta'}$$ *where α and β are roots of the equation $t^2 - 2t + 2 = 0$.*

Solution :

The roots of the equation $t^2 - 2t + 2 = 0$ are given by

$$t = \frac{1}{2}\left[2 \pm \sqrt{(4-8)}\right] = \frac{1}{2}(2 \pm 2i) = 1 \pm i$$

Let $\alpha = 1 + i$ and $\beta = 1 - i$, as α and β are the roots of the given equation

$\therefore \quad x + \alpha = x + 1 + i = (x + 1) + i$

and $x + \beta = x + 1 - i = (x + 1) - i$

Let $(x + 1) + i = r(\cos\phi + i\sin\phi)$

then $r\cos\phi = x + 1$ and $r\sin\phi = 1$

whence $\tan\phi = 1/(x + 1)$ and $r = \sqrt{[1+(x+1)^2]} = \sqrt{[1+\cot^2\phi]}$,

$\because \cot\phi = (x + 1)$

$\Rightarrow$ r cosec ϕ

Hence $x + \alpha = r(\cos\phi + i\sin\phi)$ and $x + \beta = r(\cos\phi - i\sin\phi)$,

where $r = \operatorname{cosec}\phi$ and $\tan\phi = 1/(x + 1)$

$$\therefore \quad \frac{(x+\alpha)^n - (x+\beta)^n}{\alpha - \beta} = \frac{\sin n\theta}{\sin^n\theta}$$

$$\Rightarrow \quad \frac{r^n(\cos\phi + i\sin\phi)^n - r^n(\cos\phi - i\sin\phi)^n}{(1+i)-(1-i)} = \frac{\sin n\theta}{\sin^n\theta}$$

$$\Rightarrow \quad \frac{r^n\,(2i\sin n\phi)}{2i} = \frac{\sin n\theta}{\sin^n\theta}$$

$$\Rightarrow \quad r^n \sin n\phi = \frac{\sin n\theta}{\sin^n\theta}$$

$$\Rightarrow \quad \frac{\sin n\phi}{\sin^n\phi} = \frac{\sin n\theta}{\sin^n\theta} \ldots \qquad \therefore r = \operatorname{cosec}\phi = \frac{1}{\sin\phi}$$

$$\Rightarrow \quad \phi = \theta \Rightarrow \tan\phi = \tan\theta$$

$$\Rightarrow \quad 1/(x + 1) = \tan\theta \ldots \qquad \therefore \tan\phi = 1/(x-1)$$

$$\Rightarrow \quad x + 1 = \cot\theta$$

$$\Rightarrow \quad x = \cot\theta - 1.$$

Ans.

Example 19:

If α and β be the roots of $x^2 - 2x + 4 = 0$, prove that

$$\alpha^n + \beta^n = 2^{n+1} \cos\frac{1}{3}n\pi \cos\frac{1}{3}n\pi.$$

Solution :

The roots of the equation $x^2 - 2x + 4 = 0$ are given by

$$x = \frac{2 \pm \sqrt{(4-1)}}{2} = 1 \pm 1\sqrt{3}$$

If α and β the roots of this equation, let $\alpha = 1 + \sqrt{3i}$ and $\beta = 1 - \sqrt{3i}$.

Let $1 + \sqrt{3i} = r(\cos\theta + i\sin\theta)$, then

$r\cos\theta = 1$ and $r\sin\theta = \sqrt{3}$ whence $r = 2$ and $\theta = \frac{1}{3}\pi$

$\therefore \alpha = r(\cos\theta + i\sin\theta)$ and $\beta = r(\cos\theta - \sin\theta)$

where r = 2 and $\theta = \frac{1}{3}\pi$.

$\therefore \quad \alpha^n + \beta^n = r^n (\cos\theta + i \sin\theta)^n + r^n (\cos\theta - i \sin\theta)^n$

$\Rightarrow \quad \alpha^n + \beta^n = r^n [(\cos n\theta + i \sin n\theta) + (\cos n\theta - i \sin n\theta)]$

$\Rightarrow \quad \alpha^n + \beta^n = r^n (2 \cos n\theta) = 2^n \left[2 \cos \frac{1}{3} n\pi\right] \ldots \because r = 2 \text{ and } \theta = \frac{1}{3}\pi.$

$= 2^{n+1} \cos \frac{1}{3} n\pi$ **Hence proved.**

Example 20:

Show that $(1 + i)^n = 2^{n/2} \left[\cos \frac{1}{4} n\pi + i \sin \frac{1}{4} n\pi\right]$

Hence prove that if $(1 + x)^n = p_0 + p_1 x + p_2 x^2 + p_3 x^3 + \ldots$, *then*

(i) $p_0 - p_2 + p_4 \ldots = 2^{n/2} \sin \frac{1}{4} n\pi$

(ii) $p_1 - p_3 + p_5 - \ldots = 2^{n/2} \sin \frac{1}{4} n\pi$

Solution :

Let $1 + i = r (\cos\theta + i \sin\theta)$ then

$r \cos\theta = 1$ and $r \sin\theta = 1$

whence $r = \sqrt{(1+1)} = \sqrt{2}$ and $\tan\theta = 1 \qquad \Rightarrow \theta = \frac{1}{4}\pi$

$\therefore \quad (1 + i)^n = r^n (\cos\theta + i \sin\theta)^n = r^n (\cos n\theta + i \sin n\theta)$

$= \left(\sqrt{2}\right)^n \left[\cos \frac{1}{4} n\pi + i \sin \frac{1}{4} n\pi\right] \ldots \qquad \because r = 2, \theta = \frac{1}{4}\pi$

$\therefore (1 + i)^n = 2^{n/2} \left[\cos \frac{1}{4} n\pi + i \sin \frac{1}{4} n\pi\right] \qquad \ldots(a)$

Again we are given that

$(1 + x)^n = p_0 + p_1 x + p_2 x^2 + p_3 x^3 + \ldots$

Putting x = i on both sides we get

$(1 + i)^n = p_0 + p_1 i + p_2 i^2 + p_3 i^3 + p_4 i^4 + \ldots$

$= p_0 + p_1 i - p_2 - p_3 i + p_4 + p_5 i \ldots$

$= (p_0 - p + p_4 - p_6 + \ldots) + i (p_1 - p_3 + p_5 - \ldots)$

Substituting the value of $(1 + i)^n$ from (a) we have,

$$2^{n/2}\left[\cos\frac{1}{4}n\pi + i\sin\frac{1}{4}n\pi\right] = (p_0 - p_2 + p_4 - ...) - i\,(p - p_3 + p_5 - ...)$$

Equating real and imaginary parts on both sides we have,

$$p_0 - p_2 + p_4 - ... = 2^{n/2}\cos\frac{1}{4}n\pi$$

and $p_1 - p_3 + p_5 - ... = 2^{n/2}\sin\frac{1}{4}n\pi$.

Example 21:

Evaluate *(a)* $(\cos\alpha - i\sin\alpha)^n$

(b) $(\cos\alpha + i\sin\alpha)^{-n}$

(c) $(\cos\alpha - i\sin\alpha)^{-n}$

Solution :

(a) $(\cos\alpha - i\sin\alpha)^n = [\cos(-\alpha) + \sin(-\alpha)]^n$

$= \cos(-n\alpha) + i\sin(-n\alpha)$

$= \cos n\alpha - i\sin n\alpha$ **Ans.**

(b) $(\cos\alpha + i\sin\alpha)^{-n} = \cos(-n\alpha) + \sin(-n\alpha)$

$= \cos n\alpha - i\sin n\alpha$ **Ans.**

(c) $(\cos\alpha - i\sin\alpha)^{-n} = [\cos(-\alpha) + i\sin(-\alpha)]^{-n}$

$= \cos(n\alpha) + i\sin(n\alpha)$ **Ans.**

Example 22:

Simplify $\dfrac{(\cos\alpha + i\sin\alpha)^4}{(\cos\beta + i\sin\beta)^6}$.

Solution :

$$= \frac{(\cos\alpha + i\sin\alpha)^4}{(\cos\beta + i\sin\beta)^5} = \frac{\cos 4\alpha + i\sin 4\alpha}{\cos 5\beta + i\sin 5\beta}$$

$= (\cos 4\alpha + i\sin 4\alpha)(\cos 5\beta + i\sin 5\beta)^{-1}$

$= (\cos 4\alpha + i\sin 4\alpha)[\cos(-5\beta) + i\sin(-5\beta)]$

$= \cos(4\alpha - 5\beta) + i\sin(4\alpha - 5\beta)$ **Ans.**

Example 23:

Simplify $\dfrac{(\cos\alpha + i\sin\alpha)^4}{(\sin\beta + i\cos\beta)^5}$.

Solution :

$$\frac{(\cos\alpha + i\sin\alpha)^4}{(\sin\beta + i\cos\beta)^5} = \frac{(\cos\alpha + i\sin\alpha)^4}{\left[\cos\left(\frac{1}{2}\pi - \beta\right) + i\sin\left(\frac{1}{2}\pi - \beta\right)\right]^5} \qquad \textbf{(Note)}$$

$$= \frac{\cos 4\alpha + i\sin 4\alpha}{\cos\left(\frac{5}{2}\pi - 5\beta\right) + i\sin\left(\frac{5}{2}\pi - 5\beta\right)} = \frac{\cos 4\alpha + i\sin 4\alpha}{\cos\left(\frac{1}{2}\pi - 5\beta\right) + i\sin\left(\frac{1}{2}\pi - 5\beta\right)}$$

$$\because \cos(2\pi + \theta) = \cos\theta,\ \sin(2\pi + \theta) = \sin\theta$$

$$= (\cos 4\alpha + i\sin 4\alpha)\left[\cos\left(\frac{1}{2}\pi - 5\beta\right) + i\sin\left(\frac{1}{2}\pi - 5\beta\right)\right]^{-1}$$

$$= (\cos 4\alpha + i\sin 4\alpha)\left[\cos\left(-\frac{1}{2}\pi + 5\beta\right) + i\sin\left(-\frac{1}{2}\pi + 5\beta\right)\right]$$

$$= \cos\left(4\alpha - \frac{1}{2}\pi + 5\beta\right) + i\sin\left(4\alpha - \frac{1}{2}\pi + 5\beta\right)$$

$$= \cos\left[\frac{1}{2}\pi - (4\alpha + 5\beta)\right] - i\sin\left[\frac{1}{2}\pi - (4\alpha + 5\beta)\right]$$

$$\because \cos(-\theta) = \cos\theta,\ \sin(-\theta) = -\sin\theta$$

$$= \sin(4\alpha + 5\beta) - i\cos(4\alpha + 5\beta)$$ **Ans.**

Example 24(a):

Simplify $\frac{(\cos\theta + i\sin\theta)^4}{(\cos\theta - i\sin\theta)^3}$.

Solution :

$$\frac{(\cos\theta + i\sin\theta)^4}{(\cos\theta - i\sin\theta)^3} = \frac{\cos 4\theta + i\sin 4\theta}{\cos 3\theta - i\sin 3\theta}$$

$$= (\cos 4\theta + i\sin 4\theta)(\cos 3\theta - i\sin 3\theta)^{-1}$$

$$= (\cos 4\theta + i\sin 4\theta)[\cos(-3\theta) + i\sin(-3\theta)]^{-1}$$

$$= (\cos 4\theta + i\sin 4\theta)(\cos 3\theta + i\sin 3\theta$$

$$= \cos 7\theta + i\sin 7\theta.$$ **Ans.**

Example 24(b):

Simplify $\frac{(\cos 3\theta + i\sin 3\theta)^5(\cos\theta - i\sin\theta)^3}{(\cos 5\theta + i\sin 5\theta)^7(\cos 2\theta - i\sin 2\theta)^5}$.

Solution :

$$\frac{(\cos 3\theta + i \sin 3\theta)^5 (\cos\theta - i \sin\theta)^3}{(\cos 5\theta + i \sin 5\theta)^7 (\cos 2\theta - i \sin 2\theta)^5}$$

$$= \frac{(\cos 15\theta + i \sin 15\theta)\left[\cos(-\theta) + i \sin(-\theta)\right]^3}{(\cos 35\theta + i \sin 35\theta)\left[\cos(-2\theta) + i \sin(-2\theta)\right]^5}$$

$$= \frac{(\cos 15\theta + i \sin 15\theta)\left[\cos(-3\theta) + i \sin(-3\theta)\right]}{(\cos 35\theta + i \sin 35\theta)\left[\cos(-10\theta) + i \sin(-10\theta)\right]^5}$$

$$= \frac{\cos(15\theta - 3\theta) + i \sin(15\theta - 3\theta)}{\cos(35\theta - 10\theta) + i \sin(35\theta - 10\theta)}$$

$$= \frac{(\cos 12\theta + i \sin 12\theta)}{(\cos 25\theta + i \sin 25\theta)}$$

$= (\cos 12\theta + i \sin 12\theta)(\cos 25\theta + i \sin 25\theta)^{-1}$

$= \cos(12\theta - 12\theta)[\cos(-25\theta) + i \sin(-25\theta)]$

$= \cos(12\theta - 25\theta) + i \sin(12\theta - 25\theta)$

$= \cos(-13\theta) + i \sin(-13\theta) = (\cos 13\theta - i \sin 13\theta)$. **Ans.**

Example 25(a):

If $2\cos\theta = x + \dfrac{1}{x}$, $2\cos\phi = y + \dfrac{1}{y}$, *prove that*

(i) $2\cos(\theta + \phi) = xy + \dfrac{1}{xy}$ *(ii)* $2\cos(\theta - \phi) = \dfrac{x}{y} + \dfrac{y}{x}$

(iii) $x^m y^n + 1/(x^m y^n) = 2\cos(m\theta + n\phi)$.

(iv) $(x^m/y^n) + (y^n/x^m) = 2\cos(m\theta - n\theta)$.

Solution :

(i) $xy = (\cos\theta + i \sin\theta)(\cos\phi + i \sin\phi)$

$\Rightarrow xy = \cos(\theta + \phi) + i \sin(\theta + \phi)$ ...(1)

And $1/(xy) = [\cos(\theta + \phi) + i \sin(\theta + \phi)]^{-1}$

$\Rightarrow \quad 1/(xy) = \cos(\theta + \phi) - i \sin(\theta + \phi)$ ...(2)

Adding (1) and (2) we get $xy + 1/(xy) = 2\cos(\theta + \phi)$.

(ii) $x/y = (\cos\theta + i \sin\theta)(\cos\theta + i \sin\phi)^{-1}$ **(Note)**

$= (\cos\theta + i \sin\theta)[\cos(-\phi) + i \sin(-\phi)]$

$\Rightarrow \quad x/y = \cos(\theta - \phi) + i \sin(\theta - \phi)$ ...(3)

And $y/x = 1/(x/y) = [\cos(\theta - \phi) + i \sin(\theta - \phi)]^{-1}$

$\Rightarrow \quad y/x = \cos(\theta - \phi) - i \sin(\theta - \phi)$ ...(4)

Adding (3) and (4) we get $(x/y) + (y/x) = 2\cos(\theta - \phi)$.

(iii) $x^m = (\cos\theta + i\sin\theta)^m = \cos m\theta + i \sin m\theta$

$x^n = (\cos\phi + i\sin\phi)^n = \cos n\phi + i \sin n\phi$

$\therefore \quad x^m y^n = (\cos m\theta + i\sin m\theta)(\cos n\phi + i \sin n\phi)$

$\Rightarrow \quad x^m y^n = \cos(m\theta + n\phi) + i\sin(m\theta + n\phi)$ ...(i)

Also $1/(x^m y^n) = [\cos(m\theta + n\phi) + i \sin(m\theta + n\phi)]^{-1}$

$\Rightarrow \quad 1/(x^m y^n) = \cos(m\theta + n\phi) - i \sin(m\theta + n\phi)$

...(ii)

Adding (i) and (ii) we get

$$x^m y^n + (1.x^m y^n) = 2\cos(m\theta + n\phi)$$

(iv) $x^m/y^n = (\cos m\theta + i \sin m\theta)(\cos n\phi + i\sin n\phi)^{-1}$

$= (\cos m\theta + i\sin m\theta)[\cos(-n\phi) + i \sin(-n\phi)]$

$\Rightarrow \quad x^m/y^n = \cos(m\theta - n\phi) + i \sin(m\theta - n\phi)$ ...(iii)

Also $y^n/x^m = (x^m/y^n)^{-1} = [\cos(m\theta - n\phi) + i\sin(m\theta - n\phi)]^{-1}$

$\Rightarrow \quad y^n/x^m = \cos(m\theta - n\phi) - i\sin(m\theta - n\phi)$ ...(iv)

Adding (iii) and (iv) we get $(x^m/y^n) + (y^n/x^m) = 2\cos(m\theta - n\phi)$.

Example 25(b):

If a, b, c, d denote $\cos 2\alpha + i \sin 2\alpha$, $\cos 2\beta + i \sin 2\beta$, $\cos 2\gamma + i \sin 2\gamma$ and $\cos 2\delta + i \sin 2\delta$ respectively, show that

(i) $\sqrt{(abcd)} + \dfrac{1}{\sqrt{(abcd)}} = 2\cos(\alpha + \beta + \gamma + \delta)$;

(ii) $\sqrt{\left(\dfrac{ab}{cd}\right)} + \sqrt{\left(\dfrac{cd}{ab}\right)} = 2\cos(\alpha + \beta - \gamma - \delta)$

and (iii) $\sqrt{(a^p b^q c^r)} + \dfrac{1}{\sqrt{(a^p b^b c^r)}} = 2\cos(p\alpha + q\beta + r\gamma)$.

Solution :

(i) $abcd = (\cos 2\alpha + i \sin 2\alpha)(\cos 2\beta + i\sin 2\beta)$

$\times (\cos 2\gamma + i\sin 2\gamma)(\cos 2\delta + i \sin 2\delta)$

$$= \cos(2\alpha + 2\beta + 2\gamma + 2\delta) + i\sin(2\alpha + 2\beta + 2\gamma + 2\delta)$$

$$\therefore \quad \sqrt{(abcd)} = (abcd)^{1/2} [\cos(2\alpha + 2\beta + 2\gamma + 2\delta) + i\sin(2\alpha + 2\beta + 2\gamma + 2\delta)]^{1/2}$$

$$= \cos(\alpha + \beta + \gamma + \delta) + i\sin(\alpha + \beta + \gamma + \delta)$$

and $$\frac{1}{\sqrt{(abcd)}} = (abcd)^{-1/2} = [\cos(2\alpha + 2\beta + 2\gamma + 2\delta) + i\sin(2\alpha + 2\beta + 2\gamma + 2\delta)]^{-1/2}$$

$$= \cos(\alpha + \beta + \gamma + \delta) - i\sin(\alpha + \beta + \gamma + \delta)$$

$$\therefore \quad \sqrt{(abcd)} + \frac{1}{\sqrt{(abcd)}}$$

$$= [\cos(\alpha + \beta + \gamma + \delta) + i\sin(\alpha + \beta + \gamma + \delta)] + (\cos(\alpha + \beta + \gamma + \delta) - i\sin(\alpha + \beta + \gamma + \delta)]$$

$$= \cos(\alpha + \beta + \gamma + \delta)$$ **Hence proved.**

(ii) $$\frac{ab}{cd} = \frac{(\cos 2\alpha + i\sin 2\alpha)(\cos 2\beta + i\sin 2\beta)}{(\cos 2\gamma + i\sin 2\gamma)(\cos 2\delta + i\sin 2\delta)}$$

$$= \frac{\cos(2\alpha + 2\beta) + i\sin(2\alpha + 2\beta)}{\cos(2\gamma + 2\delta) + i\sin(2\gamma + 2\delta)}$$

$$= [\cos(2\alpha + 2\beta) + i\sin(2\alpha + 2\beta)]\,[\cos(2\gamma + 2\delta) + i\sin(2\gamma + 2\delta)]^{-1}$$

$$= \cos(2\alpha + 2\beta - 2\gamma - 2\delta) + i\sin(2\alpha + 2\beta - 2\gamma - 2\delta)$$

$$\Rightarrow \sqrt{\left(\frac{ab}{cd}\right)} = \left(\frac{ab}{cd}\right)^{1/2} = [\cos 2(\alpha + \beta - \gamma - \delta) + i\sin 2(\alpha + \beta - \gamma - \delta)]^{-1/2}$$

$$= \cos(\alpha + \beta - \gamma - \delta) + i\sin(\alpha + \beta - \gamma - \delta)$$

Similarly $$\sqrt{\left(\frac{cd}{ab}\right)} = \left(\frac{ab}{cd}\right)^{-1/2}$$

$$= [\cos 2(\alpha + \beta - \gamma - \delta) + i\sin 2(\alpha + \beta - \gamma - \delta)]^{-1/2}$$

$$= \cos(\alpha + \beta - \gamma - \delta) - i\sin(\alpha + \beta - \gamma - \delta)$$

$$\therefore \quad \sqrt{\left(\frac{ab}{cd}\right)} + \sqrt{\left(\frac{cd}{ab}\right)} = [\cos(\alpha + \beta - \gamma - \delta) + i\sin(\alpha + \beta - \gamma - \delta)] + [\cos(\alpha + \beta - \gamma - \delta) - i\sin(\alpha + \beta - \gamma - \delta)]$$

$$= \cos(\alpha + \beta - \gamma - \delta)$$ **Hence proved.**

(iii) $a^p = (\cos 2\alpha + i\sin 2\alpha)^p = \cos 2p\alpha + i\sin 2p\alpha$

Similarly $b^q = \cos q\beta + i \sin 2q\beta$; $c^r = \cos 2r\gamma + i \sin 2r\gamma$

$\therefore\ a^p b^q c^r = (\cos 2p\alpha + i \sin 2p\alpha)(\cos 2q\beta + i \sin 2q\beta)$

$\times (\cos 2r\gamma + i \sin 2r\gamma)$

$= \cos (2\, p\alpha + 2q\beta + 2r\gamma) + i \sin (2p\alpha + 2q\beta + 2r\gamma)$

$= \cos 2\,(p\alpha + q\beta + r\gamma) + i \sin 2\,(p\alpha + q\beta + r\gamma)$

$\therefore\ \sqrt{(a^p b^q c^r)} = [\cos 2\,(p\alpha + 2q\beta + r\gamma) + i \sin 2\,(p\alpha + q\beta + r\gamma)]^{1/2}$

$= \cos (p\alpha + q\beta + r\gamma) + i \sin (p\alpha + q\beta + r\gamma)$

Similarly $\sqrt{\left(\frac{1}{a^p b^q c^r}\right)} = \cos (p\alpha + q\beta + r\gamma) - i \sin (p\alpha + q\beta + r\gamma)$

$\therefore\ \sqrt{(a^p b^q c^r)} + \sqrt{\frac{1}{(a^p b^q c^r)}} = [\cos (p\alpha + q\beta + r\gamma) + i \sin (p\alpha + q\beta +$

$r\gamma)]$

$+ [\cos (p\alpha + q\beta + r\gamma) - i \sin (p\alpha + q\beta + r\gamma)$

$= 2 \cos (p\alpha + qb + r\gamma)$ **Hence proved.**

Example 25(c):

Show that

$[(\cos \theta + \cos \phi) + i (\sin \theta + \sin \phi)]^n + [(\cos \theta + \cos \phi)$

$- i (\sin \theta + \sin \phi)]^n$

$= 2^{n-1} \cos^n \frac{1}{2} (\theta - \phi) \cos \frac{1}{2} n (\theta + \phi).$

Solution :

L.H.S. $= [(\cos \theta + \cos \phi) + i (\sin \theta + \sin \phi)]^n$

$+ [(\cos \theta + \cos \phi) - i (\sin \theta + \sin \phi)]^n$

$= [(2 \cos \frac{1}{2}(\theta + \phi) \cos \frac{1}{2}(\theta - \phi)\} + i \{2 \sin \frac{1}{2}(\theta+\phi) \cos \frac{1}{2}(\theta - \phi)\}]^n$

$+ [\{2 \cos \frac{1}{2}(\theta + \phi) \cos \frac{1}{2}(\theta - \phi)\} - i (2 \sin \frac{1}{2}(\theta +\phi) \cos \frac{1}{2}(\theta - \phi)\}]^n$

$= [2 \cos \frac{1}{2}(\theta - \phi) \{\cos \frac{1}{2}(\theta + \phi) + i \sin \frac{1}{2}(\theta + \phi)]^n$

$+ [2 \cos \frac{1}{2}(\theta - \phi) \{\cos \frac{1}{2}(\theta + \phi) - i \sin \frac{1}{2}(\theta + \phi)\}]^n$

$$= 2^n \cos^n \frac{1}{2}(\theta - \phi)\left[\cos\frac{1}{2}(\theta + \phi) + i \sin\frac{1}{2}(\theta + \phi)\right]^n$$

$$+ 2^n \cos^n \frac{1}{2}(\theta + \phi)\left[\cos\frac{1}{2}(\theta + \phi) - i \sin\frac{1}{2}(\theta + \phi)\right]^n$$

$$= 2^n \cos^n \frac{1}{2}(\theta - \phi)\left[\left\{\cos\frac{1}{2}n(\theta - \phi) + i \sin\frac{1}{2}n(\theta + \phi)\right\}\right.$$

$$\left.+ \left\{\cos\frac{1}{2}n(\theta - \phi) - i \sin\frac{1}{2}n(\theta + \phi)\right\}\right]$$

$$= 2^n \cos^n \frac{1}{2}(\theta - \phi)\left[2\cos\frac{1}{2}n(\theta + \phi)\right]$$

$$= 2^{n+1} \cos^n \frac{1}{2}(\theta - \phi)\cos\frac{1}{2}n(\theta + \phi) = \text{R.H.S.}$$

TO FIND THE θ ROOTS OF (COS θ – IN SIN θ)$^{P/\theta}$, WHERE P AND θ ARE INTEGERS PRIME TO EACH OTHER

In we have shown that $\cos\frac{p\theta}{q} + i \sin\frac{p\theta}{q}$ is one of the values of $(\cos\theta + \sin\theta)^{p/q}$. Here in this article we shall find out all the q different values of $(\cos\theta + i\sin\theta)^{p/q}$.

∴ We know that, $\cos(2n\pi + \theta) = \cos\theta$ and $\sin(2n\pi + \theta) = \sin\theta$, where n is any integer.

$$\therefore \cos\theta + i\sin\theta = \cos(2n\pi + \theta) + i\sin(2n\pi + \theta)$$

$$\Rightarrow (\cos\theta + i\sin\theta)^{p/q} = [\cos(2n\pi + \theta) + i\sin(2n\pi + \theta)]^{p/q}$$

$$= \cos\left\{\frac{p(2n\pi + \theta)}{q}\right\} + i\sin\left\{\frac{p(2n\pi + \theta)}{q}\right\}$$

Giving n the successive values of 0, 1, 2, 3,..., (q –1) we obtain the q values as

$$\cos\frac{p\theta}{q} + i\sin\frac{p\theta}{q}, \qquad \text{when } n = 0$$

$$\cos\frac{p(2\pi + \theta)}{q} + i\sin\frac{p(2\pi + \theta)}{q} \qquad \text{when } n = 1$$

$$\cos\frac{p(4\pi + \theta)}{q} + i\sin\frac{p(4\pi + \theta)}{q} \qquad \text{when } n = 2$$

$$\cos\frac{p}{q}\{2(\theta - 1)\pi + \theta\} + i\sin\frac{p}{q}\{2(q - 1)\pi + \theta\}, \text{ when } n = q-1$$

and these are the q different values of $(\cos\theta + i\sin\theta)^{p/q}$

Note: The highest value that should be given to n is (q – 1).

If we give n the integral values greater than (q – 1) then we shall not get any new values but same values as given by

$$n = 0, 1, 2, 3.......$$

For example let n = q, then we get the corresponding value as

$$\cos \frac{p(2q\pi+\theta)}{q} + i \sin \frac{p(2q\pi+\theta)}{q}.$$

$$= \cos p\left(2\pi + \frac{6}{q}\right) + i \sin p\left(2\pi + \frac{\theta}{q}\right)$$

$$= \cos\left(2p\pi + \frac{p\theta}{q}\right) + i \sin\left(2p\pi + \frac{p\theta}{q}\right)$$

$$= \cos \frac{p\theta}{q} + i \sin \frac{p\theta}{q},...$$

which is the same value as obtained by putting n = 0.

Example 1:

Find the cube roots of (1 – cos ϕ – i sin ϕ), where ϕ i real.

Solution :

Let 1 – cos ϕ – i sin ϕ = r (cos ϕ – i sin ϕ)

Equating real and imaginary parts we have

$$r \cos \theta = 1 - \cos \phi \text{ and } r \sin \theta = \sin \phi$$

whence $$r = \sqrt{\left\{(1-\text{co}\ \phi)^2 + \sin^2\phi\right\}} = \sqrt{(2-2\cos\phi)}$$

$$= \sqrt{\{2(1-\cos\phi)\}} = \sqrt{\left\{2\left(2\sin^2\frac{1}{2}\phi\right)\right\}} = 2\sin\frac{1}{2}\phi \text{ and}$$

$$\tan\phi = \frac{\sin\phi}{1-\cos\phi} = \frac{2\sin\frac{1}{2}\phi\cos\frac{1}{2}\phi}{2\sin^2\frac{1}{2}\phi} = \cot\frac{1}{2}\phi = \tan\left(\frac{1}{2}\pi - \frac{1}{2}\phi\right)$$

$$\Rightarrow \quad \theta = \frac{1}{2}\pi - \frac{1}{2}\phi \quad ...(ii)$$

Now $(1 - \cos\phi - i\sin\phi)^{1/3} = r^{1/3}(\cos\phi - i\sin\theta)^{1/3}$

$$= r^{1/3}[\cos(2n\pi+\theta) - i\sin(2n\pi+\theta)]^{1/3}$$

$$= r^{1/3}\left[\cos\left(\frac{2n\pi+\theta}{3}\right) - i\sin\left(\frac{2n\pi+\theta}{3}\right)\right]$$

$$= \left(2\sin\frac{\phi}{2}\right)^{1/3}\left[\cos\left\{\frac{2n\pi+\frac{1}{2}\pi-\frac{1}{2}\phi}{3}\right\} - \sin\left(\frac{2n\pi+\frac{1}{2}\pi-\frac{1}{2}\phi}{3}\right)\right]$$

putting the values of r and θ from (i) and (ii)

$$= \left(2\sin\frac{\phi}{2}\right)^{1/3}\left[\cos\left\{\frac{4n\pi+\pi-\phi}{6}\right\} - \sin\left(\frac{4n\pi+\pi-\phi}{6}\right)\right]$$

∴ The required cube roots of $(1 - \cos\phi - i\sin\phi)$ are given by

$$\left(2\sin\frac{\phi}{2}\right)^{1/3}\left[\cos\left\{\frac{4n\pi+\pi-\phi}{6}\right\} - \sin\left(\frac{4n\pi+\pi-\phi}{6}\right)\right]$$

where n = 0, 1, 2.

Example 2:

Find the five fifth roots of unity (or solve the equation $x^5 = 1$) and prove that sum of their nth power always vanishes unless n be a multiple of 5, n being an integer and that then the sum is 5.

Solution :

We know $1 = \cos 2n\pi + i\sin 2n\pi$

then $(1)^{1/5} = (\cos 2n\pi + i\sin 2n\pi)^{1/5} = \cos\frac{2n\pi}{5} + i\sin\frac{2n\pi}{5}$

Giving n the values 0, 1, 2, 3 and 4 we have the five 5th roots of unity

as $(\cos 0 + i\sin 0)$, $\left(\cos\frac{2\pi}{5} + i\sin\frac{2\pi}{5}\right)$,

$\left(\cos\frac{4\pi}{5} + i\sin\frac{4\pi}{5}\right)$, $\left(\cos\frac{6\pi}{5} + i\sin\frac{6\pi}{5}\right)$ and $\left(\cos\frac{8\pi}{5} + i\sin\frac{8\pi}{5}\right)$

i.e., 1, x, x^2, x^3 and x^5,

where $x = \cos\ \cos\frac{2\pi}{5} + i\sin\frac{2\pi}{5}$.

Sum of the nth powers of these roots

$$= 1 + x^n + x^{2n} + x^{3n} + x^{4n}.$$

$$= \frac{1-(x^n)^5}{1-x^n} \ldots \therefore 1 + r + r^2 + \ldots \text{to n terms} = \frac{1-r^n}{1-r}$$

Now $x = \cos\ \cos\dfrac{2\pi}{5} + i \sin\dfrac{2\pi}{5}$, $\because\ x^{5n} = \left(\cos\dfrac{2\pi}{5} + i\sin\dfrac{2\pi}{5}\right)^{5n}$

$\Rightarrow\ x^{5n} = \cos 2n\pi + i \sin 2n\pi = 1$

Also $x^n = \left(\cos\dfrac{2\pi}{5} + i\sin\dfrac{2\pi}{5}\right)^n = \cos\dfrac{2n\pi}{5} + i \sin\dfrac{2n\pi}{5}$

which is equal to 1, only if n be a multiple of 5, otherwise not.

$\therefore$ If n is not a multiple of 5, x^n is not equal to one and in this case the sum of the nth powers of the roots

$$= \frac{1-x^{5n}}{1-x^n} = \frac{0}{1-x^n} = 0, \text{ as } 1 - x^n \neq 0$$

Again if n is a multiple of 5, then n = 5k (say)

$$\therefore\quad x^n = x^{5k} = \left(\cos\frac{2\pi}{5} + i\sin\frac{2\pi}{5}\right)^{5k} = \cos 2k\pi + i \sin 2k\pi = 1$$

Similarly $x^{2n} = (x^n)^2 = 1$, $x^{3n} = 1 = x^{4n}$

$\therefore$ If n be a multiple of 5, then the sum of the nth powers of the roots $= 1 + x^n + x^{2n} + x^{3n} + x^{3n} + x^{4n} = 1 + 1 + 1 + 1 + 1 = 5$. **Hence proved.**

Example 3:

If sin α + sin β + sin γ = 0 cos α + cos β + cos γ, show that

(i) sin 2α + sin 2β + sin 2γ = 0 = cos 2α + cos 2β + cog 2γ

(ii) $\sin^2\alpha + \sin^2\beta + \sin^2\gamma = \dfrac{3}{2} = \cos^2\alpha + \cos^2\beta + \cos^2\gamma.$

Solution :

(i) Let $a = \cos\alpha + i\sin\alpha$, $b = \cos\beta + i\sin\beta$,

$c = \cos\gamma + i\sin\gamma$

Then $\dfrac{1}{\alpha} = \dfrac{1}{\cos\alpha + i\sin\alpha} = (\cos\alpha + i\sin\alpha)^{-1} = \cos\alpha - i\sin\alpha$

Similarly $1/b = \cos\beta - i\sin\beta$ and $1/c = \cos\gamma - i\sin\gamma$

$\therefore\ \dfrac{1}{a} + \dfrac{1}{b} + \dfrac{1}{c} = (\cos\alpha - i\sin\alpha) + (\cos\beta - i\sin\beta) + (\cos\gamma - i\sin\gamma)$

$= (\cos\alpha + \cos\beta + \cos\gamma) - i(\sin\alpha + \sin\beta + \sin\gamma)$

$= (0) - i(0)$, $\qquad \because\ \Sigma\sin\alpha = 0 = S\cos\alpha$

$= 0 \qquad$ ...(i)

Similarly $a + b + c = 0$ ...(ii)

(i) Now $(a + b + c)^2 = (a^2 + b^2 + c^2) + 2(ab + bc + ca)$

$= (a^2 + b^2 + c^2) + 2abc\,(1/c + 1/a + 1/b)$

$\Rightarrow\ 0 = (a^2 + b^2 + c^2) + 2abc\,(0)$, from (i) and (ii)

$\Rightarrow\ a^2 + b^2 + c^2 = 0$

$\Rightarrow\ (\cos\alpha + i\sin\alpha)^2 + (\cos\beta + i\sin\beta)^2 + (\cos\gamma + i\sin\gamma)^2 = 0$

$\Rightarrow\ (\cos 2\alpha + i\sin 2\alpha) + (\cos 2\beta + i\sin 2\beta) + (\cos 2\gamma + i\sin 2\gamma) =$
0

$\Rightarrow\ (\cos 2\alpha + \cos 2\beta + \cos 2\gamma) + i(\sin 2\alpha + \sin 2\beta + \sin 2\gamma) = 0$

Equating real and imaginary parts on both sides, we get

$\cos 2\alpha + \cos 2\beta + \cos 2\gamma = 0 = \sin 2\alpha + \sin 2\beta + \sin 2\gamma$

Hence proved.

(ii) If $\cos 2\alpha + \cos 2\beta + \cos 2\gamma = 0$

then $(2\cos^2\alpha - 1) + (2\cos^2\beta - 1) + (2\cos^2\gamma - 1) = 0$

$\Rightarrow\ 2(\cos^2\alpha + \cos^2\beta + \cos^2\gamma) = 3$

$\Rightarrow\ \cos^2\alpha + \cos^2\beta + \cos^2\gamma = \dfrac{3}{2}$ **Hence proved.**

Again $\cos 2\alpha + \cos 2\beta + \cos 2\gamma = 0$

$\Rightarrow\ (1 - 2\sin^2\alpha) + (1 - 2\sin^2\beta) + (1 - 2\sin^2\gamma) = 0$

$\Rightarrow\ 3 = 2(\sin^2\alpha + \sin^2\beta + \sin^2\gamma)$.

$\Rightarrow\ \sin^2\alpha + \sin^2\beta + \sin^2\gamma = \dfrac{3}{2}$. **Hence proved.**

Example 4:

Express $2 - \sqrt{3} + i$ in the form $r(\cos\theta + i\sin\theta)$.

Solution :

Let $\left(2 - \sqrt{3}\right) + i = r(\cos\theta + i\sin\theta)$ ...(i)

Since $x \neq 0$ so the given equation can be written as

$$\left(\frac{x-1}{x}\right)^n = 1 = \cos 2r\pi + i\sin 2r\pi \qquad \textbf{(Note)}$$

$$\Rightarrow\ \left(1 - \frac{1}{x}\right) = (\cos 2r\pi + i\sin 2r\pi)^{1/n}$$

$$\Rightarrow\ 1 - \frac{1}{x} = \cos\left(\frac{2r\pi}{n}\right) + i\sin\left(\frac{2r\pi}{n}\right), \text{ where } r = 0, 1, 2, \ldots, (n-1)$$

$$\Rightarrow \quad \frac{1}{x} = \left\{1 - \cos\left(\frac{2r\pi}{n}\right)\right\} - i\left\{\sin\left(\frac{2r\pi}{n}\right)\right\} \qquad \textbf{(Note)}$$

$$= 2 \sin^2 (r\pi/n) - i \{2 \sin (r\pi/n) \cos (r\pi/n)\}$$

$$= 2 \sin (r\pi/n) \{\sin (r\pi/n) - i \cos (r\pi/n)\}$$

$$= \frac{2 \sin(r\pi/n)\left\{\sin^2(r\pi/n) - i^2 \cos^2(r\pi/n)\right\}}{\sin(r\pi/n) + i \cos(r\pi/n)}, \qquad \textbf{(Note)}$$

multiplying num. and denom. by $\sin\left(\frac{r\pi}{n}\right) + i\cos\left(\frac{r\pi}{n}\right)$

$$\Rightarrow \quad \frac{1}{x} = \frac{2 \sin(r\pi/n)\{1\}}{\sin(r\pi/n) + i\cos(r\pi/n)}, \qquad \because\ i^2 = -1$$

$$\Rightarrow \quad x = \frac{\sin(r\pi/n) + i\cos(r\pi/n)}{2\sin(r\pi/n)} = \frac{1}{2}\left[1 + i\cot\left(\frac{r\pi}{n}\right)\right],$$

where $r = 0, 1, 2, \ldots, (n - 1)$. **Ans.**

Example 5:

Find all the values of $(1)^{1/4}$.

Solution :

Let $1 = r (\cos\theta + i \sin\theta)$

then $r \cos\theta = 1$ and $r \sin\theta = 0$ whence $r = 1$ and $\theta = 0$.

$$\therefore\ (1)^{1/4} = (\cos 0 + i \sin 0)^{1/4} = [\cos (2n\pi + 0) + i \sin (2n\pi + 0)]^{1/4}$$

$$= \cos (2n\pi/4) + i \sin (2n\pi/4)$$

Giving n the values 0, 1, 2 and 3 the required values are

$\cos 0 + i \sin 0$; $\cos\frac{\pi}{2} + i\sin\frac{\pi}{2}$, $\cos\pi + i\sin\pi$ and $\cos\frac{3\pi}{2} + i\sin\frac{3\pi}{2}$

i.e., $1 + i\,0$; $0 + i\,1$; $-1 + i\,0$ and $0 + i\,(-1)$

i.e., 1; i; -1, and $-i$ *i.e.,* ± 1 and $\pm i$. **Ans.**

Example6:

Find all the values of $(-1)^{1/3}$.

Solution :

$-1 = \cos\pi + i\sin\pi\ldots \qquad \because \cos\pi = -1$ and $\sin\pi = 0$

2 $(-1)^{1/3} = (\cos\pi + i\sin\pi)^{1/3}$

$= [\cos(2n\pi + \pi) + i\sin(2n\pi + \pi)]^{1/3}$

$= \cos\left\{\frac{1}{3}(2n\pi+\pi)\right\} + i\sin\left\{\frac{1}{3}(2n\pi+\pi)\right\}$

Giving n the values 0, 1 and 2, the required values are

$\cos\frac{1}{3}\pi + i\sin\frac{1}{3}\pi$, $\cos\pi + i\sin\pi$ and $\cos\frac{5}{3}\pi + i\sin\frac{5}{3}\pi$

i.e., $\frac{1}{2} + i\frac{1}{2}\sqrt{3}$, $-1 + i(0)$ and $\frac{1}{2} + i\left(-\frac{1}{2}\sqrt{3}\right)$

i.e., $\frac{1}{2}\left(1+i\sqrt{3}\right)$, -1, and $\frac{1}{2}\left(1-i\sqrt{3}\right)$ **Ans.**

Example 7:

Find all the values of $(i)^{1/4}$

Solution :

We can write $i = \cos\frac{1}{2}\pi + i\sin\frac{1}{2}\pi$ ..., $\because \cos\frac{1}{2}\pi = 0$ and $\sin\frac{1}{2}\pi = 1$

$\therefore\ (i)^{1/4} = \left(\cos\frac{1}{2}\pi + i\sin\frac{1}{2}\pi\right)^{1/4}$

$= \left[\cos\left(2n\pi + \frac{1}{2}\pi\right) + i\sin\left(2n\pi + \frac{1}{2}\pi\right)\right]^{1/4}$

$= \left[\cos\frac{1}{2}(4n\pi+\pi) + i\sin\frac{1}{2}(4n\pi+\pi)\right]^{1/4}$

$= \cos\frac{1}{8}(4n\pi + \pi) + i\sin\frac{1}{8}(4n\pi + \pi)$

Giving n the values 0, 1, 2 and 3 we have the required values as :

$\cos\frac{1}{8}\pi + i\sin\frac{1}{8}\pi$; $\cos\frac{5}{8}\pi + i\sin\frac{5}{8}\pi$;

$\cos\frac{9}{8}\pi + i\sin\frac{9}{8}\pi$ and $\cos\frac{13}{8}\pi + i\sin\frac{13}{8}\pi$

Also $\cos\frac{9}{8}\pi + i\sin\frac{9}{8}\pi = \cos\left(\pi+\frac{1}{8}\pi\right) + i\sin\left(\pi+\frac{1}{8}\pi\right)$

$= -\cos\frac{1}{8}\pi - i\sin\frac{1}{8}\pi$

And similarly $\cos\frac{13}{8}\pi + i\sin\frac{13}{8}\pi = -\cos\frac{5}{8}\pi - i\sin\frac{5}{8}\pi$

$\therefore$ The required values are

$\cos\frac{r\pi}{8} + i\sin\frac{r\pi}{8}$ and $-\cos\frac{r\pi}{8} - i\sin\frac{r\pi}{8}$, where r = 1, 5. **Ans.**

Example 8:

Find all the values of $(8i)^{1/3}$.

Solution :

$$\therefore \quad 8i = 8\left[\cos\frac{1}{2}\pi + i\sin\frac{1}{2}\pi\right]$$

$$\therefore \quad \cos\frac{1}{2}\pi = 0 \text{ and } \sin\frac{1}{2}\pi = 1$$

$$\therefore \quad (8i)^{1/3} = 8^{1/3}\left[\cos\frac{1}{2}\pi + i\sin\frac{1}{2}\pi\right]^{1/3}$$

$$= 2\left[\cos\left(2n\pi + \frac{1}{2}\pi\right) + i\sin\left(2n\pi + \frac{1}{2}\pi\right)\right]^{1/3}$$

$$= 2\left[\cos\frac{4n\pi+\pi}{6} + i\sin\frac{4n\pi+\pi}{6}\right]$$

Giving n the values 0, 1 and 2 the required values are

$$2\left[\cos\frac{1}{6}\pi + i\sin\frac{1}{6}\pi\right],\ 2\left[\cos\frac{5}{6}\pi + i\sin\frac{5}{6}\pi\right],$$

and $2\left[\cos\frac{3}{2}\pi + i\sin\frac{5}{2}\pi\right]$

i.e., $2\left[\frac{1}{2}\sqrt{3} + i.\frac{1}{2}\right], 2\left[-\frac{1}{2}\sqrt{3} + i.\frac{1}{2}\right]$ and $2\{0 + i(-1)\}$

i.e., $2.\frac{1}{2}\left(\sqrt{3}+i\right), 2.\frac{1}{2}\left(-\sqrt{3}+i\right)$ and $-2i$

i.e., $\sqrt{3+i}, -\sqrt{3+i}$ and $-2i$ **Ans.**

Example 9:

Find all the values of $(32)^{1/5}$.

Solution :

$\because \quad 32 = 32\,(1) = 32\,(\cos 0 + i \sin 0)$

$\therefore \quad (32)^{1/5} = (32)^{1/5}\,(\cos 0 + i \sin 0)^{1/5}$

$= 2\,[\cos (2n\pi + 0) + i \sin (2n\pi + 0)]^{1/5}$

$= 2\left[\cos \dfrac{2n\pi}{5} + i \sin \dfrac{2n\pi}{5}\right].$

Giving n the values 0, 1, 2, 3 and 4 the required values are

$$2\,[\cos 0 + i \sin 0],\ 2\left[\cos \frac{2\pi}{5} + i \sin \frac{2\pi}{5}\right],\ 2\left[\cos \frac{4\pi}{5} + i \sin \frac{4\pi}{5}\right]$$

$$2\left[\cos \frac{6\pi}{5} + i \sin \frac{6\pi}{5}\right] \text{ and } 2\left[\cos \frac{8\pi}{5} + i \sin \frac{8\pi}{5}\right]$$

i.e., $2\,[1 + i.0],\ 2\left[\cos \dfrac{2\pi}{5} + i \sin \dfrac{2\pi}{5}\right],\ 2\left[\cos \dfrac{4\pi}{5} + i \sin \dfrac{4\pi}{5}\right],$

$$2\left[\cos\left(2\pi - \frac{4\pi}{5}\right) + i \sin\left(2\pi - \frac{4\pi}{5}\right)\right] \text{ and}$$

$$2\left[\cos\left(2\pi - \frac{2\pi}{5}\right) + i \sin\left(2\pi - \frac{2\pi}{5}\right)\right]$$

i.e., $2\,(1),\ 2\left[\cos \dfrac{2\pi}{5} + i \sin \dfrac{2\pi}{5}\right],\ 2\left[\cos \dfrac{4\pi}{5} + i \sin \dfrac{4\pi}{5}\right],$

$$2\left[\cos \frac{4\pi}{5} - i \sin \frac{4\pi}{5}\right] \text{ and } 2\left[\cos \frac{2\pi}{5} - i \sin \frac{2\pi}{5}\right]$$

i.e., $2,\ 2\left[\cos \dfrac{2\pi}{5} \pm i \sin \dfrac{2\pi}{5}\right]$ and $\left[\cos \dfrac{4\pi}{5} \pm i \sin \dfrac{4\pi}{5}\right].$ **Ans.**

Example 10:

Find all the values of $(1 + i)^{1/3}$.

Solution :

Let $1 + i = r\,(\cos \theta + i \sin \theta)$

then $r \cos \theta = 1$ and $r \sin \theta = 1$,

whence $\quad r = \sqrt{2},\ \theta = \pi/4$

$$\therefore \quad 1 + i = \sqrt{2}\left(\cos\frac{\pi}{4} + i\sin\frac{\pi}{4}\right)$$

$$\Rightarrow \quad (2 + i)^{1/3} = \left(\sqrt{2}\right)^{1/3}\left(\cos\frac{\pi}{4} + i\sin\frac{\pi}{4}\right)^{1/3}$$

$$\Rightarrow \quad (1 + i)^{1/3} = 2^{1/6}\left[\cos\left(2n\pi + \frac{\pi}{4}\right) + i\sin\left(2n\pi + \frac{\pi}{4}\right)\right]^{1/3}$$

$$= 2^{1/6}\left[\cos\left(\frac{8n\pi + \pi}{12}\right) + i\sin\left(\frac{8n\pi + \pi}{12}\right)\right]$$

Giving n the values 0, 1 and 2 the required values are

$$2^{1/6}\left[\cos\frac{\pi}{12} + i\sin\frac{\pi}{12}\right],\ 2^{1/6}\left[\cos\frac{3\pi}{4} + i\sin\frac{3\pi}{4}\right]$$

and $2^{1/6}\left[\cos\frac{17\pi}{12} + i\sin\frac{17\pi}{12}\right]$

i.e., $2^{1/6}\left[\cos\frac{\pi}{12} + i\sin\frac{\pi}{12}\right],\ 2^{1/6}\left[-\frac{1}{\sqrt{2}} + i\frac{1}{\sqrt{2}}\right]$

and $2^{1/0}\left[\cos\frac{17\pi}{12} + i\sin\frac{17\pi}{12}\right]$

i.e., $2^{1/6}\left[\cos\frac{\pi}{12} + i\sin\frac{\pi}{12}\right],\ 2^{-/3}(-1 + i)$

and $2^{1/6}\left[\cos\frac{17\pi}{12} + i\sin\frac{17\pi}{12}\right]$

i.e., $2^{-1/3}(-1 + i)$ and $2^{1/6}\left[\cos\frac{r\pi}{12} + i\sin\frac{r\pi}{12}\right]$,

where r = 1, 17. **Ans.**

Example 11:

Prove that n, nth roots of unity form a series in G.P.

Solution :

We can write $1 = \cos 0 + i\sin 0 = \cos 2r\pi + i\sin 2r\pi$

$$\therefore \quad (1)^{1/n} = (\cos 2r\pi + i\sin 2r\pi)^{1/n}$$

$$= \cos\frac{2r\pi}{n} + i\sin\frac{2r\pi}{n}$$

Giving r the values 0, 1, 2,..., (n – 1), nth roots of unity are :

$$(\cos 0 + i \sin 0), \left(\cos \frac{2\pi}{n} + i \sin \frac{2\pi}{n}\right), \left(\cos \frac{4\pi}{n} + i \sin \frac{4\pi}{n}\right),$$

$$....., \left[\cos \frac{2(n-1)\pi}{n} + i \sin \frac{2(n-1)\pi}{n}\right]$$

i.e., $1, \left[\cos \frac{2\pi}{n} + i \sin \frac{2\pi}{n}\right], \left[\cos \frac{4\pi}{n} + i \sin \frac{4\pi}{n}\right],$

$$\left[\cos \frac{6\pi}{n} + i \sin \frac{6\pi}{n}\right], ..., \left[\cos \frac{2(n-1)\pi}{n} + i \sin \frac{2(n-1)\pi}{n}\right]$$

Let $\cos \frac{2\pi}{n} + i \sin \frac{2\pi}{n} = x$

then $x^2 = \left(\cos \frac{2\pi}{n} + i \sin \frac{2\pi}{n}\right)^2 = \cos \frac{4\pi}{n} + i \sin \frac{4\pi}{n}$,

$$x^3 = \left(\cos \frac{2\pi}{n} + i \sin \frac{2\pi}{n}\right)^3 = \cos \frac{6\pi}{n} + i \sin \frac{6\pi}{n},$$

$$x^{n-1} = \left(\cos \frac{2\pi}{n} + i \sin \frac{2\pi}{n}\right)^{n-1}$$

$$= \left[\cos \frac{2(n-1)\pi}{n} + i \sin \frac{2(n-1)\pi}{n}\right]$$

Hence from (i) the n, nth roots of unity are

$1, x, x^2, x^3, ..., x^{n-1}$, where $x = \cos \frac{2\pi}{n} + i \sin \frac{2\pi}{n}$.

These roots form G.P. whose common ratio is x

i.e., $\cos \frac{2\pi}{n} + i \sin \frac{2\pi}{n}$. **Hence proved.**

Example 12:

Show that the product of the n, nth roots of unity is $(-1)^{n-1}$.

Solution :

As in above, we can find that the n, nth roots of unity are $1, x, x^2, x^3,...,$ x^{n-1}, where $x = \cos \frac{2\pi}{n} + i \sin \frac{2\pi}{n}$.

The required product of the roots

$= 1.x.x^2.x^3...x^{n-1}$

$= x^{0+1+2....+(n-1)} = x^{(n-1)/2}$, $\quad \because \Sigma n = \frac{1}{2} n (n + 1)$

$= \left[\cos \frac{2\pi}{n} + i \sin \frac{2\pi}{n}\right]^{(n-1)n/2}$ $\quad \because x = \cos \frac{2\pi}{n} + i \sin \frac{2\pi}{n}$

$= \cos \left\{\frac{(n-1) n}{2} . \frac{2\pi}{n}\right\} + i \sin \left\{\frac{(n-1) n}{2} . \frac{2\pi}{n}\right\}$

$= \cos \{(n - 1) \pi\} + i \sin \{(n - 1) \pi\}$

$= (-1)^{n-1} + i (0)$, $\quad \because \cos n\pi = (-1)^n, \sin n\pi = 0$

$= (-1)^{n-1}$. **Hence proved.**

Example 13:

Solve the equation $x^7 + 1 = 0$.

Solution :

$x^7 + 1 = 0 \Rightarrow x^7 = -1 \Rightarrow x = (-1)^{1/7}$

Hence we are to find out all the values of $(-1)^{1/7}$

Now $-1 = \cos \pi + i \sin \pi$

$\Rightarrow (-1)^{1/7} = (\cos \pi + i \sin \pi)^{1/7}$

$= [\cos (2n\pi + \pi) + i \sin (2n\pi + \pi)]^{1/7}$

$= \cos \frac{1}{7} (2n\pi + \pi) + i \sin \frac{1}{7} (2n\pi + \pi)$

Giving n the values 0, 1, 2, 3, 4, 5 and 6 we get the required values as

$$\left(\cos \frac{\pi}{7} + i \sin \frac{\pi}{7}\right), \left(\cos \frac{3\pi}{7} + i \sin \frac{3\pi}{7}\right), \left(\cos \frac{5\pi}{7} + i \sin \frac{5\pi}{7}\right),$$

$$\left(\cos \frac{7\pi}{7} + i \sin \frac{7\pi}{7}\right), \left(\cos \frac{9\pi}{7} + i \sin \frac{9\pi}{7}\right), \left(\cos \frac{11\pi}{7} + i \sin \frac{11\pi}{7}\right)$$

$$\text{and } \left(\cos \frac{13\pi}{7} + i \sin \frac{13\pi}{7}\right) \quad ...(i)$$

Out of these values

$$\cos \frac{7\pi}{7} + i \sin \frac{7\pi}{7} = \cos \pi + i \sin \pi = -1$$

$$\cos\frac{9\pi}{7} + i\sin\frac{9\pi}{7} = \cos\left(2\pi - \frac{5\pi}{7}\right) + i\sin\left(2\pi - \frac{5\pi}{7}\right)$$

$$= \cos\frac{5\pi}{7} - i\sin\frac{5\pi}{7};$$

$$\cos\frac{11\pi}{7} + i\sin\frac{11\pi}{7} = \cos\left(2\pi - \frac{3\pi}{7}\right) + i\sin\left(2\pi - \frac{3\pi}{7}\right)$$

$$= \cos\frac{3\pi}{7} - i\sin\frac{3\pi}{7};$$

$$\cos\frac{13\pi}{7} + i\sin\frac{3\pi}{7} = \cos\left(2\pi - \frac{\pi}{7}\right) + i\sin\left(2\pi - \frac{\pi}{7}\right)$$

$$= \cos\frac{\pi}{7} - i\sin\frac{\pi}{7}.$$

Hence from (i) the required values of $(-1)^{1/7}$ or the roots of the given equation are

$$-1, \left(\cos\frac{\pi}{7} \pm i\sin\frac{\pi}{7}\right), \left(\cos\frac{3\pi}{7} \pm i\sin\frac{3\pi}{7}\right), \left(\cos\frac{5\pi}{7} \pm i\sin\frac{5\pi}{7}\right)$$

i.e., $-1, \cos\frac{r\pi}{7} \pm i\sin\frac{r\pi}{7}$, where $r = 1, 3, 5$. **Ans.**

Example 14:

Solve the equation $x^7 = 2^7$.

Solution :

$x^7 = 2^7 \Rightarrow x = (2^7)^{1/7} = 2\,(1)^7$ ...(i)

Now $1 = \cos 0 + i\sin 0 \Rightarrow (1)^{1/7} = [\cos 0 + i\sin 0]^{1/7}$

$$\Rightarrow (1)^{1/7} = [\cos(2n\pi) + i\sin(2n\pi)]^{1/7} = \cos\left(\frac{2}{7}n\pi\right) + i\sin\left(\frac{2}{7}n\pi\right)$$

Giving n the values 0, 1, 2, 3, 4, 5 and 6 we have the seven values of $(1)^{1/7}$ as

$$(\cos 0 + i\sin 0), \left(\cos\frac{2}{7}\pi + i\sin\frac{2}{7}\pi\right), \left(\cos\frac{4}{7}\pi + i\sin\frac{4}{7}\pi\right),$$

$$\left(\cos\frac{6}{7}\pi + i\sin\frac{6}{7}\pi\right), \left(\cos\frac{8}{7}\pi + i\sin\frac{8}{7}\pi\right), \left(\cos\frac{10}{7}\pi + i\sin\frac{10}{7}\pi\right)$$

and $\left(\cos\frac{12}{7}\pi + i\sin\frac{12}{7}\pi\right)$ ...(ii)

Also $\cos\frac{8}{7}\pi + i\sin\frac{8}{7}\pi = \cos\left(2\pi - \frac{6}{7}\pi\right) + i\sin\left(2\pi - \frac{6}{7}\pi\right)$

$$= \cos\frac{6}{7}\pi - i\sin\frac{6}{7}\pi$$

Similarly $\cos\frac{10}{7}\pi + i\sin\frac{10}{7}\pi = \cos\frac{4}{7}\pi - i\sin\frac{4}{7}\pi$

and $\cos\frac{12}{7}\pi + i\sin\frac{12}{7}\pi = \cos\frac{2}{7}\pi - i\sin\frac{2}{7}\pi$

∴ From (ii), the values of $(1)^{1/7}$ are

$$1, \left(\cos\frac{2}{7}\pi \pm i\sin\frac{2}{7}\pi\right), \left(\cos\frac{4}{7}\pi \pm i\sin\frac{4}{7}\pi\right), \left(\cos\frac{6}{7}\pi \pm i\sin\frac{6}{7}\pi\right)$$

Hence from (i), the roots of the given equation are2, 2 $\left(\cos\frac{2}{7}\pi \pm i\sin\frac{2}{7}\pi\right)$; 2 $\left(\cos\frac{4}{7}\pi \pm i\sin\frac{4}{7}\pi\right)$; 2 $\left(\cos\frac{6}{7}\pi \pm i\sin\frac{6}{7}\pi\right)$

$\Rightarrow$ 2, 2 $\left(\cos\frac{2r\pi}{7} \pm i\sin\frac{2r\pi}{7}\right)$, where r = 1, 2, 3.

Example 15:

Solve the equation $x^{12} - 1 = 0$ and find which of its roots satisfy the equation $x^4 + x^2 + 1 = 0$.

Solution :

$x^{12} - 1 = 0 \Rightarrow (x^6 - 1)(x^6 + 1) = 0$ ...(i)

Now $x^6 + 1 = 0$ gives $x^6 = -1 \Rightarrow x = (-1)^{1/6}$

$\Rightarrow$ $x = (\cos\pi + i\sin\pi)^{1/6}$ ... $\because -1 = \cos\pi + i\sin\pi$

$= [\cos(2n\pi + \pi) + \sin(2n\pi + \pi)]^{1/6}$

$$= \cos\left(\frac{2n\pi + \pi}{6}\right) + i\sin\left(\frac{2n\pi + \pi}{6}\right)$$

Giving n the values 0, 1, 2, 3, 4 and 5 we have the six roots of the equation $x^6 + 1 = 0$ as :

$$\left(\cos\frac{\pi}{6} + i\sin\frac{\pi}{6}\right), \left(\cos\frac{\pi}{2} + i\sin\frac{\pi}{2}\right), \left(\cos\frac{5\pi}{6} + i\sin\frac{5\pi}{6}\right),$$

$$\left(\cos\frac{7\pi}{6} + i\sin\frac{7\pi}{6}\right), \left(\cos\frac{3\pi}{2} + i\sin\frac{3\pi}{2}\right), \left(\cos\frac{11\pi}{6} + i\sin\frac{11\pi}{6}\right)$$

i.e., $\left(\cos\frac{\pi}{6} + i\sin\frac{\pi}{6}\right)$, i. $\left(\cos\frac{5\pi}{6} + i\sin\frac{5\pi}{6}\right)$, $\left[\cos\left(2\pi - \frac{5\pi}{6}\right)\right.$

$$\left. + i\sin\left(2\pi - \frac{5\pi}{6}\right)\right], -i \text{ and } \left[\cos\left(2\pi - \frac{\pi}{6}\right) + i\sin\left(2\pi - \frac{\pi}{6}\right)\right]$$

i.e., $\left(\cos\frac{\pi}{6} + i\sin\frac{\pi}{6}\right)$, i, $\left(\cos\frac{5\pi}{6} + i\sin\frac{5\pi}{6}\right)$,

$$\left(\cos\frac{5\pi}{6} - i\sin\frac{5\pi}{6}\right), i \text{ and } \left(\cos\frac{\pi}{6} - i\sin\frac{\pi}{6}\right)$$

i.e., $\pm$ i, $\left(\cos\frac{\pi}{6} \pm i\sin\frac{\pi}{6}\right)$ and $\left(\cos\frac{5\pi}{6} \pm i\sin\frac{5\pi}{6}\right)$ ...(ii)

Again $x^6 - 1 = 0$ gives $x^6 - 1 \Rightarrow x = (1)^{1/6}$

$\Rightarrow$ $x = [(\cos 0 + i \sin 0)]^{1/6}$, $\because = \cos 0 + i \sin 0$

$= [\cos (2n\pi) + i \sin (2n\pi)]^{1/6} = \cos (2n\pi/\pi) + i \sin (2n\pi/6)$

Giving n the values 0, 1, 2, 3, 4 and 5 we have the six roots of the equation $x^6 - 1 = 0$ as

$$(\cos 0 + i\sin 0), \left(\cos\frac{\pi}{3} + i\sin\frac{\pi}{3}\right), \left(\cos\frac{2\pi}{3} + i\sin\frac{2\pi}{3}\right),$$

$$(\cos \pi + i\sin \pi), \left(\cos\frac{4\pi}{3} + i\sin\frac{4\pi}{3}\right) \text{ and } \left(\cos\frac{5\pi}{3} + i\sin\frac{5\pi}{3}\right)$$

i.e., 1, $\left(\cos\frac{\pi}{3} + i\sin\frac{\pi}{3}\right), \left(\cos\frac{2\pi}{3} + i\sin\frac{2\pi}{3}\right)$, – 1,

$$\left[\cos\left(2\pi - \frac{2\pi}{3}\right) + i\sin\left(2\pi - \frac{2\pi}{3}\right)\right]$$

$$\text{and } \left[\cos\left(2\pi - \frac{\pi}{3}\right) + i\sin\left(2\pi - \frac{\pi}{3}\right)\right]$$

i.e., $1, \left(\cos\frac{\pi}{3} + i\sin\frac{\pi}{3}\right), \left(\cos\frac{2\pi}{3} + i\sin\frac{2\pi}{3}\right), -1,$

$$\left(\cos\frac{2\pi}{3} - i\sin\frac{2\pi}{3}\right) \text{ and } \left(\cos\frac{\pi}{3} - i\sin\frac{\pi}{3}\right)$$

i.e., $\pm 1, \left(\cos\frac{\pi}{3} \pm i\sin\frac{\pi}{3}\right)$ and $\left(\cos\frac{2\pi}{3} \pm i\sin\frac{2\pi}{3}\right)$ ...(iii)

Also $x^6 - 1 = (x^2 - 1)(x^4 + x^2 + 1)$.

$\therefore \quad x^2 - 1 = 0$ gives $x = \pm 1$

$\therefore$ From (iii) the roots of $x^4 + x^2 + 1 = 0$ are

$\cos\frac{\pi}{3} \pm i\sin\frac{\pi}{3}$ and $\cos\frac{2\pi}{3} \pm i\sin\frac{2\pi}{3}$

i.e., $\cos\frac{\pi}{3} \pm i\sin\frac{\pi}{3}$ and $\cos i\left(\pi - \frac{\pi}{3}\right) \pm i\sin\left(\pi - \frac{\pi}{3}\right)$

i.e., $\cos\frac{\pi}{3} \pm i\sin\frac{\pi}{3}$ and $-\left(\cos\frac{\pi}{3} \pm i\sin\frac{\pi}{3}\right)$

i.e., $\pm\left(\cos\frac{\pi}{3} \pm i\sin\frac{\pi}{3}\right)$

Hence all the roots of the equation $x^{12} - 1 = 0$ from (ii) and (iii) are

$\pm 1, \pm i, \left(\cos\frac{\pi}{6} \pm i\sin\frac{\pi}{6}\right), \left(\cos\frac{5\pi}{6} \pm i\sin\frac{5\pi}{6}\right),$

$$\left(\cos\frac{\pi}{3} \pm i\sin\frac{\pi}{3}\right) \text{ and } \left(\cos\frac{2\pi}{3} \pm i\sin\frac{2\pi}{3}\right)$$

i.e., $\pm 1, + i, \left(\cos\frac{\pi}{6} \pm i\sin\frac{\pi}{6}\right), \left[\cos\left(\pi - \frac{\pi}{6}\right) \pm i\sin\left(\pi - \frac{\pi}{6}\right)\right],$

$\left(\cos\frac{\pi}{3} \pm i\sin\frac{\pi}{3}\right)$ and $\left[\cos\left(\pi - \frac{\pi}{3}\right) \pm i\sin\left(\pi - \frac{\pi}{3}\right)\right],$

i.e., $\pm 1, \pm i, \left(\cos\frac{\pi}{6} \pm i\sin\frac{\pi}{6}\right), -\left[\cos\frac{\pi}{6} \pm i\sin\frac{\pi}{6}\right],$

$\left(\cos\frac{\pi}{3} \pm i\sin\frac{\pi}{3}\right)$ and $-\left(\cos\frac{\pi}{3} \pm i\sin\frac{\pi}{3}\right)$

i.e., $\pm 1, \pm i, \pm \left(\cos\frac{\pi}{6} \pm i \sin\frac{\pi}{6}\right)$ and $\pm\left(\cos\frac{\pi}{3} \pm i \sin\frac{\pi}{3}\right)$ **Ans.**

Example 16:

Solve the equation $x^7 + x^4 + x^3 + 1 = 0$.

Solution :

$x^7 + x^4 + x^3 + 1 = 0 \Rightarrow x^4 (x^3 + 1) + (x^3 + 1) = 0$

$\Rightarrow \quad (x^4 + 1)(x^3 + 1) = 0$ ...(i)

Then equation $x^4 + 1 = 0$ gives $x^4 = -1 \Rightarrow x = (-1)^{1/4}$

$\Rightarrow x = (\cos \pi + i \sin \pi)^{1/4} = [\cos (2n\pi + \pi + i \sin (2n\pi + \pi)]^{1/4}$

$$= \cos\left(\frac{2n\pi + \pi}{4}\right) + i \sin\left(\frac{2n\pi + \pi}{4}\right)$$

Giving n the values 0, 1, 2 and 3 the roots of the equation $x^4 + 1 = 0$ are :

$$\left(\cos\frac{\pi}{4} + i \sin\frac{\pi}{4}\right), \left(\cos\frac{3\pi}{4} + i \sin\frac{3\pi}{4}\right), \left(\cos\frac{5\pi}{4} + i \sin\frac{5\pi}{4}\right)$$

and $\left(\cos\frac{7\pi}{4} + i \sin\frac{7\pi}{4}\right)$

i.e., $\left(\frac{1}{\sqrt{2}} + i\frac{1}{\sqrt{2}}\right), \left(\frac{-1}{\sqrt{2}} + i\frac{1}{\sqrt{2}}\right), \left[\left(\frac{-1}{\sqrt{2}} + i\frac{-1}{\sqrt{2}}\right)\right]$

and $\left[\left(\frac{1}{\sqrt{2}} + i\frac{-1}{\sqrt{2}}\right)\right]$

i.e., $\frac{1+i}{\sqrt{2}}, \frac{-1-i}{\sqrt{2}}, \frac{-1+i}{\sqrt{2}}$ and $\frac{1-i}{\sqrt{2}}$ *i.e.*, $\frac{1+i}{\sqrt{2}}$ and $\frac{-1+i}{\sqrt{2}}$ ...(ii)

The equation $x^3 + 1 = 0$, gives $x^3 = -1 \Rightarrow x = (-1)^{1/3}$

$\Rightarrow x = (\cos \pi + i \sin \pi)^{1/3} = [\cos (2n\pi + \pi) + i \sin (2n\pi + \pi)]^{1/3}$

$$= \cos\frac{2n\pi + \pi}{3} + i \sin\frac{2n\pi + \pi}{3}.$$

Giving n the values 0, 1 and 2 we have the roots of the equation $x^3 + 1 = 0$ as

$$\left(\cos\frac{\pi}{3} + i\sin\frac{\pi}{3}\right), (\cos\pi + i\sin\pi) \text{ and } \left(\cos\frac{5\pi}{3} + i\sin\frac{5\pi}{3}\right)$$

i.e., $\left(\frac{1}{2} + i\frac{\sqrt{3}}{2}\right)$, -1 and $\left[\frac{1}{2} + i\left(-\frac{\sqrt{3}}{2}\right)\right]$

i.e., $\frac{1 \pm i\sqrt{3}}{2}$ and -1 ...(iii)

Hence the required roots of given equation from (ii) and (iii) are $\pm\left(\frac{1 \pm i}{\sqrt{2}}\right)$, -1 and $\frac{1 \pm i\sqrt{3}}{2}$. **Ans.**

Example 17:

Solve the equation $x^9 - x^5 + x^4 - 1 = 0$.

Solution :

$$x^9 - x^5 + x^4 - 1 = 0$$

$\Rightarrow x^5(x^4 - 1) + (x^4 - 1) = 0 \Rightarrow (x^5 + 1)(x^4 - 1) = 0$ ...(i)

The equation $x^4 - 1 = 0$ gives $x^4 = 1 \Rightarrow x^2 = \pm 1$

$\Rightarrow x = \pm\sqrt{1}$ and $\pm\sqrt{-1}$ *i.e.,* ± 1 and $\pm i$. ...(ii)

The equation $x^5 + 1 = 0$ gives $x^5 = -1 \Rightarrow x = (-1)^{1/5}$

$$\Rightarrow x = (\cos\pi + i\sin\pi)^{1/5} = [\cos(2n\pi + \pi) + i\sin(2n\pi + \pi)]^{1/5}$$

$$= \cos\left(\frac{2n\pi + \pi}{5}\right) + i\sin\left(\frac{2n\pi + \pi}{5}\right)$$

Giving n the values 0, 1, 2, 3 and 4 we have the roots of the equation $x^5 + 1 = 0$ as :

$$\left(\cos\frac{\pi}{5} + i\sin\frac{\pi}{5}\right), \left(\cos\frac{3\pi}{5} + i\sin\frac{3\pi}{5}\right), (\cos\pi + i\sin\pi),$$

$$\left(\cos\frac{7\pi}{5} + i\sin\frac{7\pi}{5}\right) \text{ and } \left(\cos\frac{9\pi}{5} + i\sin\frac{9\pi}{5}\right)$$

But $\cos\frac{7\pi}{5} + i\sin\frac{7\pi}{5} = \cos\left(2\pi - \frac{3\pi}{5}\right) + i\sin\left(2\pi - \frac{2\pi}{5}\right)$

$$= \cos\frac{3\pi}{5} - i\sin\frac{3\pi}{5}$$

and $\cos\frac{9\pi}{5} + i\sin\frac{9\pi}{5} = \cos\left(2\pi - \frac{\pi}{5}\right) + i\sin\left(2\pi - \frac{\pi}{5}\right)$

$$= \cos\frac{\pi}{5} - i\sin\frac{\pi}{5}.$$

Hence the roots of the equation $x^5 + 1 = 0$ are

$$\left(\cos\frac{\pi}{5} \pm \sin\frac{\pi}{5}\right), \left(\cos\frac{3\pi}{5} \pm \sin\frac{3\pi}{5}\right) \text{ and } (\cos\pi + i\sin\pi) \qquad \text{...(iii)}$$

Hence from (ii) and (iii) the roots of the given equation are

$$\pm 1, \pm i, (\cos\pi + i\sin\pi), \left(\cos\frac{\pi}{5} \pm \sin\frac{\pi}{5}\right)$$

and $\left(\cos\frac{3\pi}{5} \pm \sin\frac{3\pi}{5}\right)$ **Ans.**

Example 18:

Solve the equation

$$x^6 + x^5 + x^4 + x^3 + x^2 \quad x + 1 = 0.$$

Solution :

Multiplying the given equation by $(x - 1)$ we have

$$(x - 1)(x^6 + x^5 + x^4 + x^3 + x^2 + x + 1) = 0 \qquad \textbf{(Note)}$$

$$\Rightarrow x^7 - 1 = 0$$

$$\Rightarrow x = (1)^{1/7} = (\cos 0 + i\sin 0)^{1/7} = (\cos 2n\pi + i\sin 2n\pi)^{1/7}$$

$$= \cos\frac{2n\pi}{7} + i\sin\frac{2n\pi}{7}$$

Giving n the values 0, 1, 2, 3, 4, 5 and 6 we have the roots of the equation

$$(\cos 0 + i\sin 0), \left(\cos\frac{2}{7}\pi + \sin\frac{2}{7}\pi\right), \left(\cos\frac{4}{7}\pi + \sin\frac{4}{7}\pi\right)$$

$$\left(\cos\frac{6}{7}\pi + \sin\frac{6}{7}\pi\right), \left(\cos\frac{8}{7}\pi + \sin\frac{8}{7}\pi\right), \left(\cos\frac{10}{7}\pi + \sin\frac{10}{7}\pi\right)$$

and $\left(\cos\frac{12}{7}\pi + \sin\frac{12}{7}\pi\right)$

But $\cos\frac{8}{7}\pi + i\sin\frac{8}{7}\pi = \cos\frac{6}{7}\pi - i\sin\frac{6}{7}\pi$

$$\cos\frac{10}{7}\pi + i\sin\frac{10}{7}\pi = \cos\frac{4}{7}\pi - i\sin\frac{4}{7}\pi$$

and $\cos\frac{12}{7}\pi + i\sin\frac{12}{7}\pi = \cos\frac{2}{7}\pi - i\sin\frac{2}{7}\pi$

Hence the roots of the equation $x^7 - 1 = 0$ are

$$1, \left(\cos\frac{2}{7}\pi \pm i\sin\frac{2}{7}\pi\right), \left(\cos\frac{4}{7}\pi \pm i\sin\frac{4}{7}\pi\right) \text{ and } \left(\cos\frac{6}{7}\pi \pm i\sin\frac{6}{7}\pi\right)$$

The root 1 corresponds to the factor $x - 1$ and the remaining six roots are those of the given equation.

Hence the roots of the given equation are

$$\left(\cos\frac{2r\pi}{7} \pm i\sin\frac{2r\pi}{7}\right), \text{ where } r = 1, 2, 3.$$ **Ans.**

Example 19:

Solve $x^5 - x^4 + x^3 - x^2 + x - 1 = 0.$

Solution :

Multiplying the given equation by $(x + 1)$ we get

$$(x + 1)(x^5 - x^4 + x^3 - x^2 + x - 1) = 0 \Rightarrow x^6 - 1 = 0$$

Now do as above.

$$1, \cos\frac{2r\pi}{6} \pm i\sin\frac{2r\pi}{6}, \text{ where } r = 1, 2.$$ **Ans.**

Example 20:

Solve the equation $x^4 - x^3 - x + 1 = 0.$

Solution :

Multiplying the given equation by $(x + 1)$, we get

$$(x + 1)(x^4 - x^3 + x^2 - x + 1) = 0$$

$$\Rightarrow \quad x^5 + 1 = 0$$

$$\Rightarrow \quad x^5 = -1 \Rightarrow x = (-1)^{1/5}$$

$$\Rightarrow x = (\cos\pi + i\sin\pi)^{1/5} = [\cos(2n\pi + \pi) + i\sin(2n\pi + \pi)]^{1/5}$$

$$= \cos\left(\frac{2n\pi + \pi}{5}\right) + i\sin\left(\frac{2n\pi + \pi}{5}\right)$$

Giving n the values 0, 1, 2, 3 and 4 we have the roots of the equation $x^5 + 1 = 0$ as

$$\left(\cos\frac{1}{5}\pi + i\sin\frac{1}{5}\pi\right), \left(\cos\frac{3}{5}\pi + i\sin\frac{3}{5}\pi\right), (\cos\pi + i\sin\pi),$$

$$\left(\cos\frac{7}{5}\pi + i\sin\frac{7}{5}\pi\right) \text{ and } \left(\cos\frac{9}{5}\pi + i\sin\frac{9}{5}\pi\right)$$

But $\cos\frac{9}{5}\pi + i\sin\frac{9}{5}\pi = \cos\left(2\pi - \frac{1}{5}\pi\right) + i\sin\left(2\pi - \frac{1}{5}\pi\right)$

$$= \cos\frac{1}{5}\pi - i\sin\frac{1}{5}\pi,$$

$$\cos\frac{7}{5}\pi + i\sin\frac{7}{5}\pi = \cos\left(2\pi - \frac{3}{5}\pi\right) + i\sin\left(2\pi - \frac{3}{5}\pi\right)$$

$$= \cos\frac{3}{5}\pi - i\sin\frac{3}{5}\pi$$

Hence the roots of the equation $x^5 + 1 = 0$ are

$$\left(\cos\frac{1}{5}\pi \pm i\sin\frac{1}{5}\pi\right), \left(\cos\frac{3}{5}\pi \pm i\sin\frac{3}{5}\pi\right) \text{ and } -1$$

But the root -1 corresponds to the factor $x + 1$,

hence the remaining roots are those of the given equation.

Hence the roots of the given equation are

$$\left(\cos\frac{1}{5}\pi \pm i\sin\frac{1}{5}\pi\right) \text{ and } \left(\cos\frac{3}{5}\pi \pm i\sin\frac{3}{5}\pi\right).$$ **Ans.**

Example 21:

Solve that equations $x^{16} - 47x^8 + 1 = 0.,$

Solution :

$$x^{16} - 47x^8 + 1 = 0$$

$$\Rightarrow x^8 = \frac{47 \pm \sqrt{(2209-4)}}{2} = \frac{47 \pm \sqrt{(2205)}}{2} = \frac{47 \pm 21\sqrt{5}}{2}$$

$$\Rightarrow x^8 = \left(94 \pm 42\sqrt{5}\right)/4 \text{ (multiplying num. and denom. by 2)}$$

$$= \frac{49 + 45 \pm 2.7\ 3\sqrt{5}}{4} = \frac{\left(7 \pm 3\sqrt{5}\right)^2}{4} = \left[\frac{7 \pm 3\sqrt{5}}{2}\right]^2$$

$$= \left[\frac{14 \pm 6\sqrt{5}}{4}\right]^2, \text{ multiplying num. and denom. by 2}$$

$$= \left[\frac{9+5 \pm 2.3\sqrt{5}}{4}\right]^2 = \left[\left(\frac{3 \pm \sqrt{5}}{2}\right)^2\right]^2 = \left(\frac{3 \pm \sqrt{5}}{2}\right)^2$$

$$= \left[\frac{6 \pm 2\sqrt{5}}{4}\right]^4 \text{, multiplying num. and denom. by 2}$$

$$= \left[\frac{5 + 1 \pm 2.\sqrt{5}.1}{4}\right]^4 = \left[\left(\frac{\sqrt{5} \pm 1}{2}\right)^2\right]^4 = \left(\frac{\sqrt{5} \pm 1}{2}\right)^8$$

$$\Rightarrow \quad x^8 = \left(\frac{\sqrt{5} \pm 1}{2}\right)^8 (1) = \left(\frac{\sqrt{5} \pm 1}{2}\right) [\cos 2n\pi \pm i \sin 2n\pi]$$

$$\Rightarrow \quad x = \left(\frac{\sqrt{5} \pm 1}{2}\right) [\cos 2n\pi \pm i \sin 2n\pi]^{\frac{1}{8}}$$

$$= \left(\frac{\sqrt{5} \pm 1}{2}\right) \left[\cos \frac{n\pi}{4} \pm i \sin \frac{n\pi}{4}\right],$$

where $n = 0, 1, 2$ and 3. **Ans.**

Example 22:

Solve the equation $x^{10} + 11x^5 - 1 = 0$.

Solution :

$$x^{10} + 11x^5 - 1 = 0$$

$$\Rightarrow \quad x^5 = \frac{-11 \pm \sqrt{(121+4)}}{2} = \frac{-11 \pm 5\sqrt{5}}{2}$$

$$\Rightarrow \quad x = \frac{-11 \pm 5\sqrt{5}}{2} \times \frac{16}{16} = \frac{-176 \pm 80\sqrt{5}}{32} = \left(\frac{\pm\sqrt{5} - 1}{2}\right)^5$$

$$\Rightarrow \quad x^5 = \left(\frac{\pm\sqrt{5} - 1}{2}\right)^5 \times 1$$ **(Note)**

$$\Rightarrow \quad x^5 = \left(\frac{\pm\sqrt{5} - 1}{4}\right)^5 (\cos 2n\pi \pm i \sin 2n\pi)$$

$$\Rightarrow \quad x = \left(\frac{\pm\sqrt{5} - 1}{2}\right) [\cos 2n\pi \pm i \sin 2n\pi]^{1/5}$$

$$= \left(\frac{\pm\sqrt{5}-1}{2}\right)\left[\cos\frac{2n\pi}{5} \pm i \sin\frac{2n\pi}{5}\right].$$

Giving n the values 0, 1 and 2 we have the required roots, except for n = 0 we have two equal roots.

Hence the roots of the given equation are given by

$$\frac{\pm\sqrt{5}-1}{2}\left[\cos\frac{2n\pi}{5} \pm i \sin\frac{2n\pi}{5}\right], \text{ where } n = 0, 1, 2.$$

Ans.

Example 23:

Solve the equation $(x - 1)^n = x^n$, where n is a positive integer.

Solution :

Since $x \neq 0$ so the given equation can be written as

$$\left(\frac{x-1}{x}\right)^n = 1 = \cos 2r\pi + i \sin 2r\pi \qquad \textbf{(Note)}$$

$$\Rightarrow \left(1-\frac{1}{x}\right) = (\cos 2r\pi + i \sin 2r\pi)^{1/n}$$

$$\Rightarrow 1-\frac{1}{x} = \cos\left(\frac{2r\pi}{n}\right) + i \sin\left(\frac{2r\pi}{n}\right), \text{ where } r = 0, 1, 2,..., (n-1)$$

$$\Rightarrow \frac{1}{x} = \left\{1-\cos\left(\frac{2r\pi}{n}\right)\right\} - i\left\{\sin\left(\frac{2r\pi}{n}\right)\right\} \qquad \textbf{(Note)}$$

$$= 2\sin^2(r\pi/n) - i\{2\sin(r\pi/n)\cos(r\pi/n)\}$$

$$= 2\sin(r\pi/n)\{\sin(r\pi/n) - i\cos(r\pi/n)\}$$

$$= \frac{2\sin(r\pi/n)\{\sin^2(r\pi/n) - i^2\cos^2(r\pi/n)\}}{\sin(r\pi/n) + i\cos(r\pi/n)}, \qquad \textbf{(Note)}$$

multiplying num. and denom. by $\sin\left(\frac{r\pi}{n}\right) + i\cos\left(\frac{r\pi}{n}\right)$

$$\Rightarrow \frac{1}{x} = \frac{2\sin(r\pi/n)\{1\}}{\sin(r\pi/n) + i\cos(r\pi/n)}, \because \quad i^2 = -1$$

$$\Rightarrow x = \frac{\sin(r\pi/n) + i\cos(r\pi/n)}{2\sin(r\pi/n)} = \frac{1}{2}\left[1 + i\cot\left(\frac{r\pi}{n}\right)\right],$$

where r = 0, 1, 2,..., (n – 1) **Ans.**

Example 24:

Show that roots of the equation

$(1 + x)^{2n} + (1 - x)^{2n} = 0$ *are given by* $\pm i \tan \dfrac{(2k-1)\pi}{4n}$,

where $k = 1, 2, 3..., n.$

Solution :

The given equation can be written as

$$\left(\frac{1+x}{1-x}\right)^{2n} = -1 = \cos(2r+1)\pi + i\sin(2r+1)\pi$$

$\Rightarrow$ $y = [\cos(2r+1)\pi + i\sin(2r+1)\pi]^{1/n}$,

where y $[(1 + x)/(1 - x)]^2$ and r = 0, 1, 2, 3,..., n – 1.

$$\therefore\ y = \cos\left(\frac{2r+1}{n}\right)\pi + i\sin\left(\frac{2r+1}{n}\right)\pi$$

$$\Rightarrow \left(\frac{1+x}{1-x}\right)^2 = \cos\left(\frac{2r+1}{n}\right)\pi + i\sin\left(\frac{2r+1}{n}\right)\pi$$

$$\Rightarrow \frac{1+x}{1-x} = \pm\left[\cos\left(\frac{2r-1}{n}\right)\pi + i\sin\left(\frac{2r+1}{n}\right)\pi\right]^{1/2}$$

$$\Rightarrow \frac{1+x}{1-x} = \pm\cos\left(\frac{2r+1}{2n}\right)\pi \pm i\sin\left(\frac{2r+1}{2n}\right)\pi$$

$$\Rightarrow \frac{(1+x)-(1-x)}{(1+x)+(1-x)} = \frac{\pm\cos\left(\frac{2r+1}{2n}\right)\pi \pm i\sin\left(\frac{2r+1}{2n}\right)\pi - 1}{\pm\cos\left(\frac{2r+1}{2n}\right)\pi \pm i\sin\left(\frac{2r+1}{2n}\right)\pi + 1}$$

$$\Rightarrow x = \frac{\left\{-1 \pm \cos\left(\frac{2r+1}{2n}\right)\pi\right\} \pm i\sin\left(\frac{2r+1}{2n}\right)\pi}{\left\{1 \pm \cos\left(\frac{2r+1}{2n}\right)\pi\right\} \pm i\sin\left(\frac{2r+1}{2n}\right)\pi}$$

$$= \frac{\left\{1 \pm \cos\left(\frac{2r+1}{2n}\right)\pi\right\} \pm i\sin\left(\frac{2r+1}{2n}\right)\pi}{\left\{1 \pm \cos\left(\frac{2r+1}{2n}\right)\pi\right\} \pm i\sin\left(\frac{2r+1}{2n}\right)\pi} \quad ...(i)$$

Taking upper signs, we get

$$x=\frac{-\left\{2\sin^2\left(\frac{2r+1}{4n}\right)\pi\right\}+i\,2\sin\left(\frac{2r+1}{4n}\right)\pi\cos\left(\frac{2r+1}{4n}\right)\pi}{\left\{2\cos^2\left(\frac{2r+1}{4n}\right)\pi\right\}+i\,2\sin\left(\frac{2r+1}{4n}\right)\pi\cos\left(\frac{2r+1}{4n}\right)\pi}$$

$$=\frac{2i\sin\left(\frac{2r+1}{4n}\right)\pi\left[\cos\left(\frac{2r+1}{4n}\right)\pi+i\sin\left(\frac{2r+1}{4n}\right)\pi\right]}{2\cos\left(\frac{2r+1}{4n}\right)\pi\left[\cos\left(\frac{2r+1}{4n}\right)\pi+i\sin\left(\frac{2r+1}{4n}\right)\pi\right]}$$

$$=i\tan\left(\frac{2r+1}{4n}\right)\pi=i\tan\left[\frac{2(k-1)+1}{4n}\right]\pi, \text{ where } r=k-1$$

$$=i\tan\left(\frac{2k-1}{4n}\right)\pi, \text{ where } k=r+1=1, 2, 3, \ldots, n \text{ as}$$

$$r=0, 1, 2, \ldots, (n-1). \qquad \ldots(a)$$

Taking lower sign from (ii) we get

$$x=\frac{-2\cos^2\left(\frac{2r+1}{4n}\right)\pi-i\,2\sin\left(\frac{2r+1}{4n}\right)\pi\cos\left(\frac{2r+1}{4n}\right)\pi}{2\sin^2\left(\frac{2r+1}{4n}\right)\pi-i\,2\sin\left(\frac{2r+1}{4n}\right)\pi\cos\left(\frac{2r+1}{4n}\right)\pi}$$

$$=\frac{-2\cos\left(\frac{2r+1}{4n}\right)\pi\left[\cos\left(\frac{2r+1}{4n}\right)\pi+i\sin\left(\frac{2r+1}{4n}\right)\pi\right]}{-2i\sin\left(\frac{2r+1}{4n}\right)\pi\left[\cos\left(\frac{2r+1}{4n}\right)\pi+i\sin\left(\frac{2r+1}{4n}\right)\pi\right]}$$

$$=\frac{1}{i}\cot\left(\frac{2r+1}{4n}\right)\pi=i\cot\left(\frac{2r+1}{4n}\right)\pi$$

$$==-\tan\left[\frac{\pi}{2}-\frac{(2r+1)\pi}{4n}\right], \text{ since } \cot\theta=\tan\left(\frac{\pi}{2}-\theta\right)$$

$$=-i\tan\left[\frac{2n-2r-1}{4n}\right]\pi=-i\tan\left[\frac{2(n-r)-1}{4n}\right]\pi$$

$$=-i\tan\left[\frac{2k-1}{4n}\right]\pi, \text{ where } k=n-r \text{ and as } r=0, 1, 2, (n-1)$$

so $k = n, (n-1), \ldots, 2, 1$

Hence from (a), (b) the required roots are $\pm\, i \tan \left[\dfrac{2k-1}{4n}\right] \pi.$

Hence proved.

Example 25:

Find the continued product of the four values of

$$\left(\cos\frac{\pi}{3} + i\sin\frac{\pi}{3}\right)^{3/4}$$

Solution :

$$\left(\cos\frac{\pi}{3} + i\sin\frac{\pi}{3}\right)^{3/4} = \left[\cos\frac{3\pi}{3} + i\sin\frac{3\pi}{3}\right]^{1/4}$$

$$= [\cos\pi + i\sin\pi]^{1/4}$$

$$= [\cos(2n\pi + \pi) + i\sin(2n\pi + \pi)]^{1/4}$$

$$= \cos\left(\frac{2n\pi + \pi}{4}\right) + i\sin\left(\frac{2n\pi + \pi}{4}\right)$$

Giving n values 0, 1, 2 and 3 we have the four values of

$$\left(\cos\frac{\pi}{3} + i\sin\frac{\pi}{3}\right)^{3/4} \text{ as } \left(\cos\frac{\pi}{4} + i\sin\frac{\pi}{4}\right), \left(\cos\frac{3\pi}{4} + i\sin\frac{3\pi}{4}\right),$$

$$\left(\cos\frac{5\pi}{4} + i\sin\frac{5\pi}{4}\right) \text{ and } \left(\cos\frac{7\pi}{4} + i\sin\frac{7\pi}{4}\right)$$

$\therefore$ Required continued product

$$= \left(\cos\frac{\pi}{4} + i\sin\frac{\pi}{4}\right)\left(\cos\frac{3\pi}{4} + i\sin\frac{3\pi}{4}\right)\left(\cos\frac{5\pi}{4} + i\sin\frac{5\pi}{4}\right)$$

$$\times \left(\cos\frac{7\pi}{4} + i\sin\frac{7\pi}{4}\right)$$

$$= \cos\left(\frac{\pi}{4} + \frac{3\pi}{4} + \frac{5\pi}{4} + \frac{7\pi}{4}\right) + i\sin\left(\frac{\pi}{4} + \frac{3\pi}{4} + \frac{5\pi}{4} + \frac{7\pi}{4}\right)$$

$$= \cos 4\pi + i\sin 4\pi = 1 + i.0$$

$\therefore \cos 4\pi = 1$ and $\sin 4\pi = 0$. **Ans.**

Example 26:

Prove that $(a + ib)^{1/n} + (a - ib)^{1/n}$ *has n real values and find those of* $\left(1 + i\sqrt{3}\right)^{1/3} + \left(1 - i\sqrt{3}\right)^{1/3}$.

Solution :

Let $a = r\cos\theta$ and $b = r\sin\theta$,

then $r\sqrt{(a^2+b^2)}$ and $\theta = \tan^{-1}(b/a)$ ...(i)

Substituting these values of a and b we have

$$(a + ib)^{1/n} + (a - ib)^{1/n}$$

$$= r^{1/n}(\cos\theta + i\sin\theta)^{1/n} + r^{1/n}(\cos\theta - i\sin\theta)^{1/n}$$

$$= r^{1/n}[\cos(2r\pi + \theta) + i\sin(2r\pi + \theta)]^{1/n}$$

$$+ r^{1/n}[\cos(2\pi + \theta) - i\sin(2r\pi + \theta)]^{1/n}$$

$$= r^{1/n}\left[\cos\left(\frac{2r\pi +}{n}\right) + \sin\left(\frac{2r\pi+\theta}{n}\right) + \cos\left(\frac{2r\pi+\theta}{n}\right) + i\sin\left(\frac{2rp+\theta}{n}\right)\right]$$

$$= 2r^{1/n}\cos\{(2r\pi + \theta)/n\}$$

$$= 2(a^2 + b^2)^{1/2n}\cos\frac{1}{n}\left(2r\pi + \tan^{-1}\frac{b}{a}\right), \text{ from (i)}$$

which is real and will give n values corresponding to the values 0, 1, 2, 3,...(n – 1) of r.

Putting $a = 1$, $b = \sqrt{3}$ and $n = 3$ we have the three required values of $(1 + i\sqrt{3})^{1/3} + (1 - i\sqrt{3})^{1/3}$ as

$$2(1 + 3)^{1/6}\cos\frac{1}{3}(2r\pi + \tan^{-1}\sqrt{3}), \text{ where } r = 0, 1, 2$$

i.e., $2(2^2)^{1/6}\cos\frac{1}{3}\left(2r\pi+\frac{\pi}{3}\right)$, where $r = 1, 2$

i.e., $2 \times 2^{1/3}\cos\frac{1}{3}\left(\frac{\pi}{3}\right), 2\times 2^{1/3}\cos\frac{1}{3}\left(\frac{7\pi}{3}\right)$

and $2 \times 2^{1/3}\cos\frac{1}{3}\left(\frac{13\pi}{3}\right)$

i.e., $2^{4/3}\cos\frac{\pi}{9}, 2^{4/3}\cos\frac{7\pi}{9}$ and $2^{4/3}\cos\frac{13\pi}{9}$

i.e., $2^{4/3}\cos\frac{1}{9}(r\pi)$,

where $r = 1, 7, 13.$ **Ans.**

Example27:

Find all the values of $(-1)^{1/3}$.

Solution :

We have

$-1 = \cos\pi + i\sin\pi...$ $\qquad \because \cos\pi = -1$ and $\sin\pi = 0$.

$\therefore \quad (-1)^{1/3} = (\cos\pi + i\sin\pi)^{1/3}$

$= [\cos(2n\pi + \pi) + i\sin(2n\pi + \pi)]^{1/3}$

$$= \cos\left\{\frac{1}{3}(2n\pi+\pi)\right\} + i\sin\left\{\frac{1}{3}(2n\pi+\pi)\right\}$$

Giving n the values 0, 1 and 2, the required values are

$$\cos\frac{1}{3}\pi + i\sin\frac{1}{3}\pi,\ \cos\pi + i\sin\pi \text{ and } \cos\frac{5}{3}\pi + i\sin\frac{5}{3}\pi$$

i.e., $\frac{1}{2} + i\frac{1}{2}\sqrt{3},\ -1 + i(0)$ and $\frac{1}{2} + i\left(-\frac{1}{2}\sqrt{3}\right)$

i.e., $\frac{1}{2}\left(1+i\sqrt{3}\right),\ -1,$ and $\frac{1}{2}\left(1-i\sqrt{3}\right)$ **Ans.**

$\Rightarrow \quad \cos^2\alpha + \cos^2\beta + \cos^2\gamma = \frac{3}{2}$ **Hence proved.**

Again $\quad \cos 2\alpha + \cos 2\beta + \cos 2\gamma = 0$

$\Rightarrow \quad (1 - 2\sin^2\alpha) + (1 - 2\sin^2\beta) + (1 - 2\sin^2\gamma) = 0$

$\Rightarrow \quad 3 = 2(\sin^2\alpha + \sin^2\beta + \sin^2\gamma).$

$\Rightarrow \quad \sin^2\alpha + \sin^2\beta + \sin^2\gamma = \frac{3}{2}.$ **Hence proved.**

Example 28:

Prove that $(\sin x + i\cos x)^n$

$$= \cos n\left(\frac{1}{2}\pi - x\right) + i\sin n\left(\frac{1}{2}\pi - x\right).$$

Solution :

$$(\sin x + i\cos x)^n = \left[\cos\left(\frac{1}{2}\pi - x\right) + i\sin\left(\frac{1}{2}\pi - x\right)\right]^n$$

$$= \cos n\left(\frac{1}{2}\pi - x\right) + i\sin\left(\frac{1}{2}\pi - x\right)$$ **Hence proved.**

Example 29:

Express $2 - \sqrt{3} + i$ in the form $r(\cos\theta + i\sin\theta)$.

Solution :

Let $\left(2-\sqrt{3}\right) + i = r(\cos\theta + i\sin\theta)$ ...(i)

Equating real and imaginary parts on both sides we get

$2-\sqrt{3} = r\cos\theta$ ...(ii), $1 = r\sin\theta$ ...(iii)

Squaring and adding (ii) and (iii) we get

$$r^2 = \left(2-\sqrt{3}\right)^2 + (1)^2 = 4 + 3 - 4\sqrt{3} + 1 = 4\left(2-\sqrt{3}\right)$$

$$= 2\left(4-2\sqrt{3}\right) = 2\left(3+1-2\sqrt{3}\right) = 2\left(\sqrt{3}-1\right)^2$$

$$\therefore\ r = \sqrt{2}\left(\sqrt{3}-1\right) = \left(\sqrt{6}-\sqrt{2}\right) \quad \text{...(iv)}$$

Again from (ii) and (iii) we get $\tan\theta = 1/\left(2-\sqrt{3}\right)$

$$\Rightarrow \tan\theta = \frac{2+\sqrt{3}}{\left(2-\sqrt{3}\right)\left(2+\sqrt{3}\right)} = 2+\sqrt{3} \quad \text{...(v)}$$

$\therefore$ With the help of (iv) and (v) we can get the required expression from (i).

Example 30:

If $\sin\alpha + \sin\beta + \sin\gamma = 0\ \cos\alpha + \cos\beta + \cos\gamma$, *show that*

(i) $\sin 2\alpha + \sin 2\beta + \sin 2\gamma = 0 = \cos 2\alpha + \cos 2\beta + \text{cog}\ 2\gamma$

(ii) $\sin^2\alpha + \sin^2\beta + \sin^2\gamma = \dfrac{3}{2} = \cos^2\alpha + \cos^2\beta + \cos^2\gamma.$

Solution :

(i) Let $a = \cos\alpha + i\sin\alpha$, $b = \cos\beta + i\sin\beta$,

$c = \cos\gamma + i\sin\gamma$

Then $\dfrac{1}{\alpha} = \dfrac{1}{\cos\alpha + i\sin\alpha} = (\cos\alpha + i\sin\alpha)^{-1} = \cos\alpha - i\sin\alpha$

Similarly $1/b = \cos\beta - i\sin\beta$ and $1/c = \cos\gamma - i\sin\gamma$

$$\therefore\ \frac{1}{a}+\frac{1}{b}+\frac{1}{c} = (\cos\alpha - i\sin\alpha) + (\cos\beta - i\sin\beta) + (\cos\gamma - i\sin\gamma)$$

$$= (\cos\alpha + \cos\beta + \cos\gamma) - i(\sin\alpha + \sin\beta + \sin\gamma)$$

$= (0) - i(0)$, $\because \Sigma\sin\alpha = 0 = S\cos\alpha$

$= 0$ ...(i)

Similarly $a + b + c = 0$ ...(ii)

(i) Now $(a + b + c)^2 = (a^2 + b^2 + c^2) + 2(ab + bc + ca)$

$$= (a^2 + b^2 + c^2) + 2abc\,(1/c + 1/a + 1/b)$$

$\Rightarrow \quad 0 = (a^2 + b^2 + c^2) + 2abc\,(0)$, from (i) and (ii)

$$\Rightarrow \quad a^2 + b^2 + c^2 = 0$$

$$\Rightarrow \quad (\cos\alpha + i\sin\alpha)^2 + (\cos\beta + i\sin\beta)^2 + (\cos\gamma + i\sin\gamma)^2 = 0$$

$$\Rightarrow \quad (\cos 2\alpha + i\sin 2\alpha) + (\cos 2\beta + i\sin 2\beta) + (\cos 2\gamma + i\sin 2\gamma) = 0$$

$$\Rightarrow \quad (\cos 2\alpha + \cos 2\beta + \cos 2\gamma) + i\,(\sin 2\alpha + \sin 2\beta + \sin 2\gamma) = 0$$

Equating real and imaginary parts on both sides, we get

$$\cos 2\alpha + \cos 2\beta + \cos 2\gamma = 0 = \sin 2\alpha + \sin 2\beta + \sin 2\gamma$$

(ii) If $\cos 2\alpha + \cos 2\beta + \cos 2\gamma = 0$

then $(2\cos^2\alpha - 1) + (2\cos^2\beta - 1) + (2\cos^2\gamma - 1) = 0$

$$\Rightarrow \quad 2(\cos^2\alpha + \cos^2\beta + \cos^2\gamma) = 3$$

$$\Rightarrow \quad \cos^2\alpha + \cos^2\beta + \cos^2\gamma = \frac{3}{2}$$

Again $\quad \cos 2\alpha + \cos 2\beta + \cos 2\gamma = 0$

$$\Rightarrow \quad (1 - 2\sin^2\alpha) + (1 - 2\sin^2\beta) + (1 - 2\sin^2\gamma) = 0$$

$$\Rightarrow \quad 3 = 2(\sin^2\alpha + \sin^2\beta + \sin^2\gamma).$$

$$\Rightarrow \quad \sin^2\alpha + \sin^2\beta + \sin^2\gamma = \frac{3}{2}.$$

Hence proved.

2

EXPANSION OF TAN α IN POWER OF α

INTRODUCTION

(i) Expansion of sin α

$$\sin n\theta = n \cos^{n-1}\theta \sin\theta - \frac{n(n-1)(n-2)}{1.2.3}\cos^{n-3}\sin^3\theta + \ldots$$

Putting $n\theta = \alpha \Rightarrow n = \alpha/\theta$ ...(i)

we have $$\sin\alpha = \frac{\alpha}{\theta}\cos^{n-1}\theta \sin\theta - \frac{\alpha}{\theta}\frac{\left(\frac{\alpha}{\theta}-1\right)\left(\frac{\alpha}{\theta}-2\right)}{1.2.3} \times \cos^{n-3}\sin^3\theta + \ldots$$

$$= \frac{\alpha}{\theta}\cos^{n-1}\theta \sin\theta - \frac{(\alpha-\theta)(\alpha-2\theta)}{\theta^3\, 3!}\cos^{n-3}\theta\sin^3\theta + \ldots$$

$$\Rightarrow \sin\alpha = \alpha\cos^{n-1}\theta\left(\frac{\sin\theta}{\theta}\right)\frac{\alpha(\alpha-\theta)(\alpha-2\theta)}{3!}\cos^{n-3}\theta\left(\frac{\sin\theta}{\theta}\right)^3 + \ldots$$

...(ii)

Let n increase indefinitely, α remaining constant, then from (i) we find θ decreases indefinitely. Consequently (sin θ)/θ and cos θ in the limit when θ is very small are equal to unity and so also their every power is equal to unity. Hence from (ii) we have

$$\sin\alpha = \sin\alpha = \alpha - \frac{\alpha^3}{3!} + \frac{\alpha^5}{5!} - \ldots \text{ ad. inf.} \qquad \text{(Remember)} \qquad \ldots\text{(iii)}$$

Also if u_n be the nth term of this series (iii), then

$$\lim_{n\to\infty}\frac{u_{n+1}}{|u_n|} = \lim_{n\to\infty}\frac{\alpha^{2n+1}/(2n+1)!}{\alpha^{2n-1}/(2n-1)!} = \lim_{n\to\infty}\frac{\alpha^2}{(2n+1)(2n)}$$

$$= 0 < 1$$

Hence Σu_n or the series (iii) is convergent.

(ii) **Expansion of cos α:** We have proved in that

$$\cos n\theta = \cos^n \theta + \frac{n(n-1)}{1.2}\cos^{n-2}\theta \sin^2\theta + \frac{n(n-1)(n-2)(n-3)}{1.2.3.4} \times \cos^{n-4}\theta \sin^4\theta - \ldots$$

Putting $n\theta = \alpha \Rightarrow n = \alpha/\theta$

$$\cos\alpha = \cos^n\theta - \frac{\frac{\alpha}{\theta}\left(\frac{\alpha}{\theta}-1\right)}{2\,!}\cos^{n-2}\theta\sin^2\theta + \frac{\frac{\alpha}{\theta}\left(\frac{\alpha}{\theta}-1\right)\left(\frac{\alpha}{\theta}-2\right)\left(\frac{\alpha}{\theta}-3\right)}{4\,!}\cos^{n-4}\theta\sin^4\theta + \ldots$$

$$= \cos^n\theta - \frac{\alpha(\alpha-\theta)}{\theta^2\,2\,!}\cos^{n-2}\theta\sin^2\theta + \frac{\alpha(\alpha-\theta)(\alpha-2\theta)(\alpha-3\theta)}{\theta^4\,4\,!} \times \cos^{n-4}\theta\sin^4\theta - \ldots$$

$$\Rightarrow \cos\alpha = \cos^n\theta - \frac{\alpha(\alpha-\theta)}{2\,!}\cos^{n-2}\theta\left(\frac{\sin\theta}{\theta}\right)^2 + \frac{\alpha(\alpha-\theta)(\alpha-2\theta)(\alpha-3\theta)}{4\,!}\left(\frac{\sin\theta}{\theta}\right)^4 - \ldots \qquad \ldots(i)$$

As in part (i) when n increases indefinitely, α remaining constant, θ decreases indefinitely and consequently (sin θ)/θ and cos θ in the limit when θ is very small are equal to unity and so are every power of them equal to unity. Hence from (i) above we find that

$$\cos\alpha = 1 - \frac{\alpha^2}{2\,!} + \frac{\alpha^4}{4\,!} - \ldots \text{ ad. inf.} \qquad \text{(Remember)} \qquad \ldots(ii)$$

If u_n be the nth term of this series (ii) then

$$\lim_{n\to\infty}\frac{u_{n+1}}{|u_n|} = \lim_{n\to\infty}\frac{\alpha^{2n}/(2n)\,i}{\alpha^{2n-2}/(2n-2)\,!} = \lim_{n\to\infty}\frac{\alpha^2}{(2n)(2n-1)}$$

$$= 0 < 1.$$

Hence S u_n or the series (ii) is convergent.

EXPANSION OF TAN α IN POWERS OF α

$$\tan\alpha = \frac{\sin\alpha}{\cos\alpha} = \frac{\alpha - \frac{\alpha^3}{3\,!} + \frac{\alpha^5}{5\,!} - \ldots}{1 - \frac{\alpha^2}{2\,!} + \frac{\alpha^4}{4\,!} - \ldots}$$

$$=\left(\alpha-\frac{\alpha^3}{6}+\frac{\alpha^5}{10}-...\right)\left(1-\frac{\alpha^2}{2}+\frac{\alpha^4}{24}-...\right)^{-1}$$

$$=\left(\alpha-\frac{\alpha^3}{6}+\frac{\alpha^5}{120}-...\right)\left[1-\left(\frac{\alpha^2}{2}+\frac{\alpha^4}{24}-...\right)\right]^{-1}$$

$$=\left(\alpha-\frac{\alpha^3}{6}+\frac{\alpha^5}{120}-...\right)\left[1+\left(\frac{\alpha^2}{2}-\frac{\alpha^4}{24}-...\right)+\left(\frac{a^2}{2}+\frac{a^4}{24}+...\right)^2+...\right]$$

$$=\left(\alpha-\frac{\alpha^3}{6}+\frac{\alpha^5}{120}-...\right)\left(1+\frac{\alpha^2}{2}-\frac{\alpha^4}{24}+\frac{\alpha^4}{4}+...\right)$$

$$=\left(\alpha-\frac{\alpha^3}{6}+\frac{\alpha^5}{120}-...\right)\left(1+\frac{\alpha^2}{2}+\frac{5\alpha^4}{24}+...\right)$$

$$=\alpha+\frac{\alpha^3}{2}+\frac{5\alpha^5}{24}+...-\frac{\alpha^3}{6}-\frac{\alpha^5}{12}-...+\frac{\alpha^5}{120}+...$$

$$=\alpha+\alpha^3\left(\frac{1}{2}-\frac{1}{6}\right)+\alpha^5\left(\frac{1}{24}-\frac{1}{12}+\frac{1}{120}\right)+...$$

$$\Rightarrow \tan\alpha=\alpha+\frac{1}{3}\alpha^3+\frac{2}{15}\alpha^5+...\text{ ad. inf.} \qquad \text{(Remember)}$$

EXPANSION OF SIN nθ AND COS nθ

We know by De Moivre's Theorem that

$$(\cos\theta+i\sin\theta)^n=\cos n\theta+i\sin n\theta \qquad ...(i)$$

Also if n be a positive integer, then with the help of Binomial Theorem, we have

$$(\cos\theta+i\sin\theta)^n=c\,x^n\,\theta+{}^nc_1\cos^{n-1}\theta\,(i\sin\theta)+{}^nc_2\cos^{n-2}\theta \times (i\sin\theta^2+{}^nc_3\cos^{n-3}\theta\,(i\sin\theta)^3+...$$

$$=\cos^n\theta+i\,{}^nc_1\cos^{n-1}\theta\sin\theta-{}^nc_2\cos^{n-2}\theta\sin^2\theta - i\,{}^nc_3\cos^{n-3}\theta\sin^3\theta+...,\text{ since } i=\sqrt{(1)}$$

$$\Rightarrow (\cos\theta+i\sin)^n=[\cos^n\theta-{}^nc_2\cos^{n-2}\theta\sin^2\theta+{}^nc_4\cos^{n-4}\sin^4\theta-...]+i\,[{}^nc_1\cos^{n-1}\theta\sin\theta-{}^nc_3\cos^{n-3}\theta\sin^3\theta+...] \qquad ...(ii)$$

∴ From (i) and (ii) we have

$$\cos n\theta+i\sin n\theta=[\cos^n\theta-{}^nc_2\cos^{n-2}\theta\sin^2\theta+{}^nc_4\cos^{n-4}\theta\sin^4\theta-...]+i\,[{}^nc_1\cos^{n-1}\theta\sin\theta-{}^nc_3\cos^{n-3}\theta\sin^3\theta+...]$$

Equating real and imaginary parts on both sides, we have

$$\cos n\theta = \cos^n \cos\theta\ {}^nc_2 \cos^{n-2}\theta \sin^2\theta + {}^nc_4 \cos^{n-4}\theta \sin^4\theta - \ldots$$

and $\sin n\theta = {}^nc_1 \cos^{n-1}\theta \sin\theta - {}^nc_3 \cos^{n-3}\theta \sin^3\theta$

$$+ {}^nc_5 \cos^{n-5}\theta \sin^5\theta - \ldots$$

i.e., $\cos n\theta = \cos^n\theta - \frac{n(n-1)}{1.2}\cos^{n-2}\theta \sin^2\theta$

$$+ \frac{n(n-1)(n-2)(n-3)}{1.2.3.4}\cos^{n-4}\theta \sin^4\theta - \ldots$$

and $\sin n\theta = n\cos^{n-1}\theta \sin\theta - \frac{n(n-1)(n-2)}{1.2.3}\cos^{n-3}\theta \sin^3\theta$

$$+ \frac{n(n-1)(n-2)(n-3)(n-4)}{1.2.3.4.5}\cos^{n-5}\theta \sin^5\theta - \ldots$$

The terms in each of the above series are alternately positive and negative and each series continues till a factor in the numerator vanishes.

Example 1:

Expand cos 4θ.

Solution :

By De Moivre's Theorem,

$$(\cos\theta + i\sin\theta)^4 = \cos 4\theta + i\sin 4\theta \quad \ldots(i)$$

Also by Binomial Theorem, we have

$$(\cos\theta + i\sin\theta)^4 = \cos^4\theta + 4\cos^3\theta\,(i\sin\theta) + \frac{4.3}{1.2}\cos^2\theta\,(i\sin\theta)^2$$

$$+ \frac{4.3.2}{1.2.3}\cos\theta\,(i\sin\theta)^3 + \frac{4.3.2.1}{1.2.3.4}(i\sin\theta)^4$$

$$= \cos\theta + 4i\cos^2\theta\sin\theta - \theta\cos^2\theta\sin^2\theta$$

$$- 4i\cos\theta\sin^3\theta + \sin^4\theta$$

$$(\cos\theta + \sin\theta)^4 = (\cos^4\theta - 6\cos^2\theta\sin^2\theta + \sin^4\theta)$$

$$+ i(4\cos^3\theta\sin\theta - 4\cos\theta + \sin^3\theta) \quad \ldots(ii)$$

∴ From (i) and (ii), we have

$$\cos 4\theta + i\sin 4\theta = (\cos^4\theta - 6\cos^2\theta\sin^2\theta + \sin^4\theta$$

$$+ i(4\cos^3\theta\sin\theta - 4\cos\theta\sin^3\theta)$$

Equating real parts on both sides we get

$$\cos 4\theta = \cos^4\theta - 6\cos^2\theta\sin^2\theta + \sin^4\theta.$$ **Ans.**

Example 2:

Expand cos 7θ.

Solution :

By De Moivre's Theorem,

$$(\cos\theta + i\sin\theta)^7 = (\cos 7\theta + i\sin 7\theta) \quad ...(i)$$

Also by Bionomial Theorem, we have

$$(\cos\theta + i\sin\theta)^7 = \cos^7\theta + 7c_1\cos^6\theta\,(i\sin\theta) + 7c_2\cos^5\theta\,(i\sin\theta)^2 + 7c_3\cos^4\theta\,(i\sin\theta)^3 + 7c_4\cos^3\theta\,(i\sin\theta)^4 + 7c_5\cos^2\theta\,(i\sin\theta)^5 + 7c_6\cos\theta\,(i\sin\theta)^6 + 7c_7\,(i\sin\theta)^7$$

$$= \cos^7\theta + 7i\cos^6\theta\sin\theta - 21\cos^5\theta\sin^2\theta - 35i\cos^4\theta\sin^3\theta + 35\cos^3\theta\sin^4\theta + 21\,i\cos^2\theta\sin^5\theta - 7\cos\theta\sin^6\theta - i\sin^7\theta$$

$$\Rightarrow (\cos\theta + i\sin\theta)^7 = (\cos^7\theta - 21\cos^5\theta\sin^2\theta + 35\cos^3\theta\sin^4\theta - 7\cos\theta\sin^6\theta) + i\,(7\cos^6\theta\sin\theta - 35\cos^4\theta\sin^3\theta + 21\cos^2\theta\sin^5\theta - \sin^7\theta) \quad ...(ii)$$

∴ From (i) and (ii) we get

$$\cos 7\theta + i\sin 7\theta = \cos^7\theta - 21\cos^5\theta\sin^2\theta + 35\cos^3\theta\sin^4\theta - 7\cos\theta\sin^6\theta) + i\,(7\cos^6\sin\theta - 35\cos^4\theta\sin^3\theta + 21\cos^2\theta\sin^5\theta - \sin^7\theta)$$

Equating real parts on both sides we have

$$\cos 7\theta = \cos^7\theta - 21\cos^5\theta\sin^2\theta + 35\cos^3\sin^4\theta - 7\cos\theta\sin^6\theta$$

Example 3:

Expand sin 7θ.

Solution :

By De Moivre's Theorem,

$$(\cos\theta + i\sin\theta)^7 = \cos 7\theta + i\sin 7\theta \quad ...(i)$$

Also by Binomial Theorem, we have

$$(\cos\theta + i\sin\theta)^7 = \cos^7\theta + 7c_1\cos^6\theta\,(i\sin\theta) + 7c_2\cos^5\theta\,(i\sin\theta)^2 + 7c_3\cos^4\theta\,(i\sin\theta)^3 + 7c_4\cos^3\theta\,(i\sin\theta)^4 + 8c_5\cos^2\theta\,(i\sin\theta)^5 + 7c_6\cos\theta\,(i\sin\theta)^6 + 7c_7\,(i\sin\theta)^7$$

$$= \cos^7\theta + 7i\cos^6\theta\sin\theta - 21\cos^5\theta\sin^2\theta - 35i\cos^4\theta\sin^3\theta + 35\cos^3\theta\sin^4\theta + 21i\cos^2\theta\sin^5\theta\sin^5\theta - 7\cos\theta\sin^6\theta - i\sin^7\theta$$

$$= (\cos^7\theta - 21\cos^5\theta\sin\theta + 35\cos^3\theta\sin^4\theta - 7\cos\theta\sin^6\theta)$$

$+ i (7 \cos^6 \theta \sin \theta - 35 \cos^4 \theta \sin^3 \theta + 21 \cos^2 \theta \sin^5 \theta -$

$\theta - \sin^7 \theta)$ ...(ii)

∴ From (i) and (ii) we get

$\cos 7\theta + i \sin 7\theta = (\cos^7 \theta - 21 \cos^5 \theta \sin^2 \theta + + 35 \cos^3 \theta \sin^4 \theta$

$- \cos \theta \sin^6 \theta) + i (7 \cos^6 \theta \sin \theta - 35 \cos^4 \theta \sin^3 \theta$

$+ 21 \cos^2 \theta \sin^5 - \sin^7 \theta)$

Equating imaginary parts on both sides we have

$\sin 7\theta = 7 \cos^6 \theta \sin \theta - 35 \cos^4 \sin^3 \theta + 21 \cos^2 \theta \sin^5 \theta - \sin^7 \theta.$

Ans.

Example 4:

Expand sin 8θ.

Solution :

By De Moivre's Theorem.

$(\cos \theta + i \sin \theta)^8 = \cos 8\theta + i \sin 8\theta$ ...(i)

Also by Binomial Theorem, we have

$(\cos \theta + i \sin \theta)^8 = \cos^8 + 8c_1 \cos^7 \theta (i \sin \theta) + 8c_2 \cos^6 \theta (i \sin \theta)^2$

$+ 8c_3 \cos^5 \theta (i \sin \theta)^3 + 8c_4 \cos^4 \theta (i \sin \theta)^4 + 8c_5 \cos^3 \theta (i \sin \theta)^5$

$+ 8c_6 \cos^2 \theta (i \sin \theta)^6 + 8c_7 \cos \theta (i \sin \theta)^7 + 8c_8 (i \sin \theta)^8$

$= \cos^8 \theta + 8i \cos^7 \theta \sin \theta - 28 \cos^6 \theta \sin^2 \theta - 56i \cos^5 \theta \sin^3 \theta$

$+ 70 \cos^4 \theta \sin^4 \theta + 56i \cos^3 \theta \sin^5 \theta - 8 \cos^2 \theta \sin^6 \theta$

$- 8i \cos \theta \sin^7 \theta + \sin^8 \theta$

$\Rightarrow$ $(\text{cow } \theta + i \sin \theta)^8 = (\cos^8 \theta - 28 \cos^6 \sin^2 \theta + 70 \cos^4 \theta \sin^4 \theta$

$- 28 \cos^2 \theta \sin^6 \theta + \sin^8 \theta) + i [8 \cos^7 \theta \sin \theta - 56 \cos^5 \theta \sin^3 \vartheta$

$+ 56 \cos^3 \theta \sin^5 \theta - 8 \cos \theta \sin^7\theta]$ ...(ii)

∴ From (i) and (ii) we have

$\cos 8\theta + i \sin 8\theta = [\cos^8 \theta - 28 \cos^6 \theta \sin^2 \theta + 70 \cos^4 \theta \sin^4 \theta$

$- 28 \cos^2 \theta \sin^6 \theta + \sin^8 \theta]$

$+ i [8 \cos^7 \theta \sin \theta - 56 \cos^5 \theta \sin^3 \theta + 56 \cos^3 \theta \sin^5 \theta$

$- 8 \cos \theta \sin^7 \theta]$

Equating imaginary parts on both sides, we have

$\sin 8\theta = 8 \cos^7 \theta \sin \theta - 56 \cos^5 \theta \sin^3 \theta + 56 \cos^3 \theta \sin^5 \theta$

$- 8 \cos \theta \sin^7 \theta.$ **Ans.**

Example 5:

Expand (sin 9θ)/(sin θ) in a series of ascending powers of cos θ.

Solution :

By De Moivre's Theorem, we get

$$(\cos\theta + i\sin\theta)^9 = \cos 9\theta + i\sin 9\theta \qquad \text{...(i)}$$

Also by Binomial Theorem we have

$$(\cos\theta + i\sin\theta)^9 = \cos^9\theta + 9c_1\cos^8\theta\,(i\sin\theta)$$
$$+ 9c_2\cos^7\theta\,(i\sin\theta)^2 + 9c_3\cos^6\theta\,(i\sin\theta)^3 + 9c_4\cos^5\theta\,(i\sin\theta)^4$$
$$+ 9c_5\cos^4\theta\,(i\sin\theta)^5 + 9c_6\cos^3\theta\,(i\sin\theta)^6$$
$$+ 9c_7\cos^2(i\sin\theta)^7 + 9c_8\cos\theta\,(i\sin\theta)^8 + 9c_9\,(i\sin\theta)^9$$
$$= \cos^9 + 9i\cos^8\theta\sin\theta - 36\cos^7\theta\sin^2\theta - 84i\cos^6\theta\sin^3\theta$$
$$+ 126\cos^5\theta\sin^4\theta + 126i\cos^4\theta\sin^5\theta - 84\cos^3\theta\sin^6\theta$$
$$- 36i\cos^2\theta\sin^7\theta + 9\cos\theta\sin^8\theta + i\sin^9\theta$$
$$= (\cos^9\theta - 36\cos^7\theta\sin^2\theta + 126\cos^3\theta\sin^4\theta - 84\cos^3\theta\sin^6\theta$$
$$+ 9\cos\theta\sin^8\theta) + i\,(\theta\cos^8\theta\sin\theta - 84\cos^6\theta\sin^3\theta$$
$$+ 126\cos^4\theta\sin^5\theta - 36\cos^2\theta\sin^7\theta + \sin^9\theta)$$
$$\text{...(ii)}$$

∴ From (i) and (ii) we have

$$\cos 9\theta + i\sin 9\theta = (\cos^9\theta - 36\cos^7\theta\sin^2\theta + 126\cos^5\theta\sin^4\theta$$
$$- 84\cos^3\theta\sin^6\theta + 9\cos\theta\sin^8\theta) + i\,(9\cos^8\theta\sin\theta$$
$$- 84\cos^6\theta\sin^3\theta + 126\cos^4\theta\sin^5\theta - 36\cos^2\theta\sin^7\theta + \sin^9\theta)$$

Equating imaginary parts on both sides, we have

$$\sin 9\theta = 9\cos^8\theta\sin\theta - 84\cos^6\theta\sin^3\theta + 126\cos^4\theta\sin^5\theta$$
$$- 36\cos^2\theta\sin^7\theta + \sin^9\theta$$

$$\Rightarrow \frac{\sin 9\theta}{\sin\theta} = 9\cos^8\theta - 84\cos^6\theta\sin^2\theta + 126\cos^4\theta\sin^4\theta$$
$$- 36\cos^2\theta\sin^6\theta + \sin^8\theta$$
$$= 9\cos^8\theta - 84\cos^6\theta\,(1-\cos^2\theta)^3 + 126\cos^4\theta\,(1-\cos^2\theta)^2$$
$$- 36\cos^2\theta\,(1-\cos^2\theta)^3 + (1-\cos^2\theta)^4, \quad \therefore \sin^2\theta = 1-\cos^2\theta$$
$$= 9\cos^8\theta - 84\cos^6\theta + 84\cos^8\theta + 126\cos^4\theta\,(1 - 2\cos^2\theta$$
$$+ \cos^4\theta) - 36\cos^2\theta\,(1 - 3\cos^2\theta + 3\cos^4\theta - \cos^6\theta)$$
$$+ (1 - 4\cos^2\theta + 6\cos^4\theta - 4\cos^6\theta + \cos^8\theta)$$
$$= 1 - 40\cos^2\theta + 240\cos^4\theta - 448\cos^6\theta + 256\cos^8\theta.$$ **Ans.**

Example 6:

When n is odd, prove that the last terms in the expansions of sin nθ and cos nθ are respectively

$(-1)^{(n-1)/2} \sin^n \theta$ *and* $n(-1)^{(n-1)/2} \cos\theta \sin^{n-1}\theta$.

Solution :

If n is odd, then (n – 1) is even.

Now by De Moivre's Theorem

$$(\cos\theta + i\sin\theta)^n = \cos n\theta + i\sin n\theta \qquad ...(i)$$

Also by Binomial Theorem, we have

$$(\cos\theta + i\sin\theta)^n = \cos^n\theta + {}^nc_1 \cos^{n-1}\theta\,(i\sin\theta) + {}^nc_2 \cos^{n-2}\theta\,(i\sin\theta)^2 + {}^nc_3 \cos^{n-3}\theta\,(i\sin\theta)^3 ... ...+ {}^nc_{n-1} \cos\theta\,(i\sin\theta)^{n-1} + {}^nc_n (i\sin\theta)^n \theta$$

$$\Rightarrow \cos n\theta + i\sin n\theta = \cos^n\theta + {}^nc_1\, i\cos^{n-1}\theta\sin\theta - {}^nc_2 \cos^{n-2}\theta \sin^2\theta - {}^nc_3\, i\cos^{n-3}\theta\sin^3\theta + ... + {}^nc_{n-1} i^{n-1}\cos\theta\sin^{n-1}\theta + {}^nc_n\, i^n \sin^n\theta, \text{ from (i)}$$

But $i^{n-1} = (i^2)\,(n-1)^{/2} = (-1)^{(n-1)/2}$

and $i^n = i \times i^{n-1} = i\,(-1)^{(n-1)/2}$ $\qquad \because i^{n-1} = (-1)^{(n-1)/2}$

$$\therefore \cos n\theta + i\sin n\theta = \cos^n\theta + {}^nc_1\, i\cos^{n-1}\theta\sin\theta - {}^nc_2\cos^{n-2}\theta\sin^2\theta - {}^nc_3\, i\cos^{n-3}\theta\sin^3\theta + (-1)^{(n-1)/2}\, {}^nc_{n-1}\cos\theta \sin^{n-1}\theta + i\,(-1)^{(n-1)/2}\, {}^nc_n \sin^n\theta$$

$$= [\cos^n\theta - {}^nc_2\cos^{n-2}\theta\sin^2\theta + ... (-1)^{(n-1/2}\, ({}^nc_{n-1})\cos\theta\sin^{n-1}\theta] + i\,[{}^nc_1\cos^{n-1}\theta\sin\theta - {}^nc_3\cos^{n-3}\theta\sin^3\theta + ... + (-1)^{(n-1)/2}\, {}^nc_n \sin^n\theta]$$

Equating real and imaginary parts both sides, we have

$$\cos n\theta = \cos^n\theta - {}^nc_2\cos^{n-2}\theta\sin^2\theta + ... + (-1)^{(n-1)/2}\, {}^nc_{n-1} \cos\theta\sin^n\theta$$

and $\sin n\theta = {}^nc_1\cos^{n-1}\theta\sin\theta\ {}^nc_3\cos^{n-3}\theta\sin^3\theta + ... + (-1)^{(n-1)/2}\, {}^nc_{n-1}\sin^n\theta$

∴ The required last terms in the expansions of cos nθ and sin nθ are $(-1)^{(n-1)/2}\,({}^nc_{n-1})\cos\theta\sin^{n-1}\theta$ and $(-1)^{(n-1)/2}\, {}^nc_n \sin^n\theta$ respectively

i.e., $(-1)^{(n-1)/2}.n\cos\theta\sin^{n-1}\theta$ and $(-1)^{(n-1)/2}.1.\sin^n\theta$ respectively

$$\because {}^nc_{n-1} = n \text{ and } {}^nc_n = 1.$$

Example 7:

When n is even prove that the last terms in the expansions of sin nθ and cos nθ are respectively

$n(-1)^{(n-2)/2}\cos\theta\sin^{n-1}\theta$ *and* $(-1)^{n/2}\sin^n\theta$.

Solution :

As in last example we can prove that

$$\cos n\theta + i\sin n\theta = \cos^n\theta + i.{}^nc_1\cos^{n-1}\theta\sin\theta - {}^nc_2\cos^{n-2}\theta\sin^2\theta - \ldots + {}^nc_{n-1}\, i^{n-1}\cos\theta\sin^{n-1}\theta + {}^nc_n\, i^n\sin^n\theta$$

$$\Rightarrow \cos n\theta + i\sin n\theta = \cos^n\theta + i.n\cos^{n-1}\theta\sin\theta - \frac{n(n-1)}{1.2}\cos^{n-2} \ldots + ni^{n-1}\cos\theta\sin^{n-1}\theta + i^n\sin^n\theta \qquad \ldots(i)$$

Now if n is even, n – 2 is also even

Also $i^{n-1} = i \times i^{n-2} = i\,(i^2)^{(n-2)/2}$ **(Note)**

$= i\,(-1)^{(n-2)/2}$, $\because i^2 = -1$

and $i^n = (i^2)^{n/2} = (-1)^{n/2}$, $\because i^2 = -1$

∴ From (i) substituting these values of i^{n-1} and i^n, we have

$$\cos n\theta + i\sin n\theta = \cos^n\theta + i\,n\cos^{n-1}\theta\sin\theta - \frac{n(n-1)}{1.2}\cos^{n-2}\theta\sin^2\theta \ldots + ni(-1)^{(n-2)/2}\cos^n\theta\sin^{n-1}\theta + (-1)^{n/2}\sin^n\theta$$

$$= \left[\cos^n\theta - \frac{n(n-1)}{1.2}\cos^{n-2}\theta\sin^2\theta + \ldots + (-1)^{n/2}\sin^n\theta\right]$$

$$+ i\left[n\cos^{n-1}\theta\sin\theta - \frac{n(n-1)(n-2)}{1.2.3}\cos^{n-3}\theta\sin^3\theta + \ldots \ldots + n(-1)^{(n-2)/2}\cos\theta\sin^{n-1}\theta\right]$$

Equating real and imaginary parts on both sides, we get

$$\cos n\theta = \cos^n\theta - \frac{n(n-1)}{1.2}\cos^{n-2}\theta\sin^2\theta + \ldots + (-1)^{n/2}\sin^n\theta$$

$$\text{and } \sin n\theta = n\cos^{n-1}\theta\sin\theta - \frac{n(n-1)(n-2)}{1.2.3}\cos^{n-3}\theta\sin^3\theta + \ldots + n(-1)^{(n-2)/2}\cos\theta\sin^{n-1}\theta$$

Hence the required last terms in the expansions of cos nθ and sin nθ are $(-1)^{(n-2)/2}\, n\cos\theta\sin^{n-1}\theta$ respectively.

Example 8:

Prove that

$$\sin n\theta \cos^n \theta = n \tan \theta - \frac{n(n-1)(n-2)}{1.2.3} \tan^3 \theta + \ldots$$

$$\text{and } \cos n\theta \cos^n \theta = 1 - \frac{n(n+1)}{1.2} \tan^2 \theta + \frac{n(n+1)(n+2)(n+3)}{1.2.3.4} \tan^4 \theta + \ldots$$

Solution :

By De Moivre's Theorem

$$(\cos \theta + i \sin \theta)^{-n} = \cos n\theta - i \sin n\theta \qquad \ldots(i)$$

Also with the help of Binomial Theorem, we have

$(\cos \theta + i \sin \theta)^{-n} = [\cos \theta(1 + i \tan \theta)]^{-n} = (\cos \theta)^{-n} [1 + i \tan \theta]^{-n}$,

$\therefore$ the index is negative

$$= \sec^n \theta \left[1 - n(i \tan \theta) + \frac{(-n)(-n-1)}{1.2} (i \tan \theta)^2 + \frac{(-n)(-n-1)(-n-2)}{1.2.3} (i \tan \theta)^3 + \frac{(-n)(-n-1)(-n-2)(-n-3)}{1.2.3.4} (i \tan \theta)^4 + \ldots\right]$$

$$= \sec^n \theta \left[1 - n\, i \tan\theta - \frac{n(n+1)}{1.2} \tan^2 \theta + \frac{n(n+1)(n+2)}{1.2.3} i \tan^3 \theta + \frac{n(n+1)(n+2)(n+3)}{1.2.3.4} \tan^4 \theta - \ldots\right] \qquad \ldots(ii)$$

$\therefore$ From (i) and (ii), we have

$\cos n\theta - i \sin n\theta$

$$= \sec^n \theta \left[\left\{1 - \frac{n(n+1)}{1.2} \tan^2 \theta + \frac{n(n+1)(n+2)(n+3)}{1.2.3.4} \tan^4 \theta \ldots\right\} - i \left\{n \tan \theta - \frac{n(n+1)(n+2)}{1.2.3} \tan^3 \theta + \ldots\right\}\right]$$

Equating real and imaginary parts on both sides we have $\cos n\theta$

$$= \sec^n\theta\left[1-\frac{n(n+1)}{1.2}\tan^2\theta+\frac{n(n+1)(n+2)(n+3)}{1.2.3.4}\tan^4\theta\ldots\right]$$

$$\text{and } \sin n\theta = \sec^n\theta\left[n\tan\theta-\frac{n(n+1)(n+2)}{1.2.3}\tan^3\theta+\ldots\right]$$

i.e, $\cos n\theta \cos^n\theta$

$$= 1-\frac{n(n+1)}{1.2}\tan^2\theta+\frac{n(n+1)(n+2)(n+3)}{1.2.3.4}\tan^4\theta-\ldots$$

$$\text{and } \sin n\theta\cos^n\theta = n\tan\theta-\frac{n(n+1)(n+2)}{1.2.3}\tan^4\theta+\ldots$$

Hence proved.

EXPANSION OF TAN Nθ

In we have proved that

$$\sin n\theta = n\cos^{n-1}\theta\sin\theta-\frac{n(n-1)(n-2)}{1.2.3}\cos^{n-3}\theta\sin^3\theta+\ldots$$

$$\text{and } \cos n\theta = \cos^n\theta-\frac{n(n-1)}{1.2}\cos^{n-2}\sin^2\theta$$

$$+\frac{n(n-1)(n-2)(n-3)}{1.2.3.4}\cos^{n-4}\theta\sin^4\theta-\ldots$$

∴ Dividing we have

$$\tan n\theta = \frac{\sin n\theta}{\cos n\theta}$$

$$= \frac{n\cos^{n-1}\theta\sin\theta-\frac{n(n-1)(n-2)}{1.2.3}\cos^{n-3}\sin^3\theta+\ldots}{\cos^n\theta-\frac{n(n-1)}{1.2}\cos^{n-2}\sin^2\theta+\frac{(n-1)(n-2)(n-3)}{1.2.3.4}\cos^{n-4}\theta\sin^4\theta}$$

Case I. n is odd: The last terms in the expansion of sin nθ and cos nθ are respectively $(-1)^{(n-1)/2}\sin^n\theta$ and $n(-1)^{(n-1)/2}\cos\theta\sin^{n-1}\theta$.

∴ In the expansion of tan nθ, the last term in the numerator is $(-1)^{(n-1)/2}\tan^n\theta$ and in the denominator is $n(-1)^{(n-1)/2}\tan^{n-1}\theta$.

Case II. n is even: The last terms in the expansion of sin nθ and cos nθ are respectively

$$n(-1)^{(n-2)/2}\cos\theta\sin^{n-1}\theta \text{ and } (-1)^{n/2}\sin n\theta$$

∴ In the expansion of tan nθ, the last term in the numerator is $n(-1)^{(n-2/2}\tan^{n-1}\theta$ and in the denominator is $(-1)^{n/2}\tan^n\theta$.

Example 1:

Expand tan 5θ.

Solution :

We have proved above that

$$\tan n\theta = \frac{n \tan\theta - \frac{n(n-1)(n-2)}{1.2.3}\tan^3\theta + \ldots}{1 - \frac{n(n-1)}{1.2}\tan^2\theta + \frac{n(n-1)(n-2)(n-3)}{1.2.3.4}\tan^4\theta - \ldots}$$

$$\therefore \text{ Here } \tan 5\theta = \frac{5\tan\theta - \frac{54.3}{1.2.3}\tan^3\theta + \frac{5.4.3.2.1}{1.2.3.4.5}\tan^5\theta}{1 - \frac{5.4}{1.2}\tan^2\theta + \frac{5.4.3.2.1}{1.2.3.4.5}\tan^4\theta}$$

$$= \frac{5\tan\theta - 10\tan^3\theta + \tan^5\theta}{1 - 10\tan^2\theta + 5\tan^4\theta}$$ **Ans.**

Example 2:

Expand tan 8θ.

Solution :

We know tan nθ

$$= \frac{n\tan\theta - \frac{n(n-1)(n-2)}{1.2.3}\tan^3\theta + \ldots}{1 - \frac{n(n-1)}{1.2}\tan^2\theta + \frac{n(n-1)(n-2)(n-3)}{1.2.3.4}\tan^4\theta - \ldots}$$

$$\therefore \tan 8\theta = \frac{8\tan\theta - \frac{8.7.6}{1.2.3}\tan^2\theta + \frac{8.7.6.5.4}{1.2.3.4.5}\tan^5\theta - \frac{8.7.6.5.4.3.2}{1.2.3.4.5.6.7}\tan^7\theta}{1 - \frac{8.7}{1.2}\tan^2\theta + \frac{8.7.6.5}{1.2.3.4}\tan^4\theta - \frac{8.7.6.5.4.3}{1.2.3.4.5.6}\tan^6\theta + \frac{8.7.6.5.4.3.2.1}{1.2.3.4.5.6.7.8}\tan^8\theta}$$

$$= \frac{8\tan\theta - 55\tan^3\theta + 56\tan^5\theta - 8\tan^7\theta}{1\ 2\tan^2\theta + 70\tan^4\theta - 28\tan^6\theta + \tan^8\theta}$$ **Ans.**

Example 3:

Show that

$$\tan\frac{\theta}{n}+\tan\frac{\theta+\pi}{n}+\tan\frac{\theta+2\pi}{n}+\ldots+\tan\frac{\theta+(n-1)\pi}{n}$$

is equal to n tan θ or – n cot θ according as n is odd or even.

Solution :

Let α denote any one of the angles

$$\frac{\theta}{n},\frac{\theta+\pi}{n},\frac{\theta+2\pi}{n},\ldots,\frac{\theta+(n-1)\pi}{n} \quad \ldots(i)$$

Then $n\alpha=\theta+m\pi$, where $m=0, 1, 2,\ldots,(n-1)$

$\Rightarrow\ \tan n\alpha=\tan(\theta+m\pi)=\tan\theta \quad \ldots(ii)$

Case I. n is odd: Then from (ii) we get

$$\frac{{}^nc_1\tan\alpha-{}^nc_3\tan^3\alpha+\ldots+(-1)^{(n-1)/2}\tan^n\alpha}{1-{}^nc_2\tan^2\alpha+\ldots+(-1)^{(n-1)/2}\,n\tan^{n-1}\alpha}=\tan\theta \quad \textbf{(Note)}$$

$\Rightarrow\ \tan^n\alpha-n\tan\theta\tan^{n-1}\alpha+\ldots=0, \quad \ldots(iii)$

cross multiplying and simplifying

This equation, given by (iii), is an nth degree equation in tan α and from (iii) we find that the roots of the equation (ii) are

$$\tan\frac{\theta}{n},\ \tan\frac{\theta+\pi}{n},\ \tan\frac{\theta+2\pi}{n},\ldots,\tan\frac{\theta+(n-1)\pi}{n}$$

$\therefore$ Sum of the roots of (iii) is given by

$$\tan\frac{\theta}{n}+\tan\frac{\theta+\pi}{n}+\ldots+\tan\frac{\theta+(n-1)\pi}{n}=n\tan\theta$$ **Hence proved.**

Case II. n is even: The from (ii), we get

$$\frac{{}^nc_1\tan\alpha-{}^nc_3\tan^3\alpha+\ldots+(-1)^{(n-2)/2}\,n\tan^{n-1}\alpha}{1-{}^nc_2\tan^2\alpha+\ldots+(-1)^{n/2}\tan^n\alpha}$$

$\Rightarrow\ \tan^n\alpha\tan\theta+n\tan^{n-1}\alpha+\ldots=0, \quad \ldots(iv)$

cross multiplying and simplifying.

This equation given by (iv) is an nth degree equation in tan a and its roots are the same as in Case I above.

$\therefore$ Sum of the roots of (iv) is given by

$$\tan\frac{\theta}{n}+\tan\frac{\theta+2\pi}{n}+\ldots+\tan\frac{\theta+(n-1)\pi}{n}=-\frac{n}{\tan\theta}=-n\cot\theta$$

Hence proved.

EXPANSION OF TAN (A + B + Γ + Δ + ...).

Proof : We know

$(\cos\alpha + i\sin\alpha)(\cos\beta + i\sin\beta)(\cos\gamma + i\sin\gamma)(\cos\delta + i\sin\delta)...$

$= \cos(\alpha+\beta+\gamma+\delta+...) + i\sin(\alpha+\beta+\gamma+\delta+...)$ ...(i)

Now we may write $\cos\alpha + i\sin\alpha = \cos\alpha(1 + i\tan\alpha)$

Similarly $\cos\beta + i\sin\beta = \cos\beta(1 + i\tan\beta)$,

$\cos\gamma + i\sin\gamma = \cos\gamma(1 + i\tan\gamma)$, etc.

Substituting these values in (i) we have

$\cos(\alpha+\beta+\gamma+\delta+...) + i\sin(\alpha+\beta+\gamma+\delta+...)$

$= \cos\alpha(1 + i\tan\alpha)\cos\beta(1 + i\tan\beta)\,\text{cog}\,\gamma(1 + i\tan\gamma)$

$\times \cos\delta(1 + i\tan\delta)...$

$= \cos\alpha\cos\beta\cos\gamma\cos\delta...(1 + i\tan\alpha)(1 + i\tan\beta)(1 + i\tan\gamma)$

$\times (1 + i\tan\delta)...$

$= \cos\alpha\cos\beta\cos\gamma\cos\delta...[1 + i(\tan\alpha + \tan\beta + \tan\gamma + \tan\delta + ...)$

$+ i^2(\tan\alpha\tan\beta + \tan\alpha\tan\gamma + \tan\beta\tan\gamma + ...)$

$+ i^3(\tan\alpha\tan\beta\tan\gamma + \tan\beta\tan\gamma\tan\delta + ...) + ...]$

$= \cos\alpha\cos\beta\cos\gamma\cos\delta...[1 + is_1 + i^2s_2 + i^2s_3 + ...]$, ...(ii)

where $s_1 = \Sigma\tan\alpha = \tan\alpha + \tan\beta + \tan\gamma + \tan\delta + ...$,

$s_2 = \Sigma\tan\alpha\tan\beta = \tan\alpha\tan\beta + \tan\alpha\tan\gamma + \tan\beta\tan\gamma + ...$,

$s_3 = \Sigma\tan\alpha\tan\beta\tan\gamma = \tan\alpha\tan\beta\tan\gamma + \tan\beta\tan\gamma\tan\delta + ...$etc.

Hence from (ii) we get

$\cos(\alpha+\beta+\gamma+\delta+...) + i\sin(\alpha+\beta+\gamma+\delta+...)$

$= \cos\alpha\cos\beta\cos\gamma\cos\delta...[1 + is_1 - s_2 - is_3 + s_4 + is_5 - ...]$

Equating real and imaginary parts we have

$\cos(\alpha+\beta+\gamma+\delta+...) = \cos\alpha\cos\beta\cos\gamma\cos\delta...[1 - s_2 + s_4 - ...]$

and $\sin(\alpha+\beta+\gamma+\delta+...) = \cos\alpha\cos\beta\cos\gamma\cos\delta...$

$[s_1 - s_2 + s_5 + ...]$

∴ Dividing we have,

$$\tan(\alpha+\beta+\gamma+\delta+...) = \frac{\sin(\alpha+\beta+\gamma+\delta+....)}{\cos(\alpha+\beta+\gamma+\delta+....)} = \frac{s_1 - s_2 + s_5 - ...}{1 - s_2 + s_4 - ...}$$

Example 1:

If α, β, γ be the roots of the equation $x^3 + px^2 + qx + p = 0$ *prove that* $\tan^{-1}\alpha + \tan^{-1}\beta + \tan^{-1}\gamma = n\pi$ *radians except in one particular case.*

Solution :

If α, β, γ be the roots of the given equation

$x^3 + px^2 + qx + p = 0$, then we have

$$\alpha + \beta + \gamma = -p/1 = -p$$

$$\alpha\beta + \alpha\gamma + \beta\gamma = \theta/1 = \theta \text{ and } \alpha\beta\gamma = -p/1 = -p \qquad \text{...(i)}$$

Also if $\alpha = \tan\theta_1$, $\beta = \tan\theta_2$ and $\gamma = \tan\theta_3$ then

$$\tan(\theta_1 + \theta_2 + \theta_3) = \frac{"s_1 - s_3"}{1 - s_2}$$

$$= \frac{(\tan\theta_1 + \tan\theta_2 + \tan\theta_3) - (\tan\theta_1 \tan\theta_2 \tan\theta_3)}{1 - (\tan\theta_1 \tan\theta_2 + \tan\theta_1 \tan\theta_2 + \tan\theta_2 \tan\theta_3)}$$

$$= \frac{(\alpha + \beta + \gamma)}{1 - (\alpha\beta + \alpha\gamma + \beta\gamma)} = \frac{(-p) - (-p)}{1 - q}, \text{ from (i)}$$

$= 0$, except when $\theta = 1$.

$\Rightarrow \theta_1 + \theta_2 + \theta_3 = n\pi$

i.e., $\tan^{-1}\alpha + \tan^{-1}\beta + \tan^{-1}\gamma = n\pi$, except when $q = 1$.

Example 2:

If α, β, γ and δ be the roots of the equation

$x^4 - x^3 \sin 2\theta + x^2 \cos 2\theta x \cos\theta \sin\theta = 0$, *prove that*

$$\tan^{-1}\alpha + \tan^{-1}\beta + \tan^{-1}\gamma + \tan^{-1}\delta = n\pi + \frac{1}{2}\pi - \theta.$$

Solution :

If α, β, γ and δ be the roots of the equation

$x^4 - x^3 \sin 2\theta + x^2 \cos 2\theta - x\cos\theta - \sin\theta = 0$, then we have

$$\alpha + \beta + \gamma + \delta = -(\sin 2\theta)/1 = \sin 2\theta,$$

$$\alpha\beta + \alpha\gamma + \alpha\delta + \beta\gamma + \beta\delta + \gamma\delta = \cos 2\theta/1 = \cos 2\theta$$

$$\alpha\beta\gamma + \alpha\beta\delta + \alpha\gamma\delta + \beta\gamma\delta = \frac{(-\cos\theta)}{1} = \cos\theta$$

and $\alpha\beta\gamma\delta = -\sin\theta$

Let $\alpha = \tan\theta_1$, $\beta = \tan\theta_2$, $\gamma = \tan\theta_3$ and $\delta = \tan\theta_4$, then

$$\tan(\theta_1 + \theta_2 + \theta_3 + \theta_4) = \frac{"s_1 - s_3"}{1 - s_2 + s_4}$$

$$= \frac{(\Sigma \tan \theta_1) - (\Sigma \tan \theta_1 \tan \theta_2 \tan \theta_3)}{1-(\Sigma \tan \theta_1 \tan \theta_2) + (\Sigma \tan \theta_1 \tan \theta_2 \tan \theta_3 \theta_4)}$$

$$= \frac{(\Sigma\alpha)\ (\Sigma\alpha\beta\gamma)}{1-\Sigma\alpha\beta+\Sigma\alpha\beta\gamma} \qquad \because \tan \theta_1 = a \text{ etc.}$$

$$= \frac{(\alpha+\beta+\gamma+\delta) - (\alpha\beta\gamma+\alpha\beta\delta+\alpha\gamma\delta+\beta\gamma\delta)}{1-(\alpha\beta+\alpha\gamma+\alpha\delta+\beta\gamma+\beta\delta+\gamma\delta)+\alpha\beta\gamma\delta}$$

$$= \frac{\sin 2\theta - \cos\theta}{1-\cos 2\theta - \sin\theta}, \text{ substituting the values from (i)}$$

$$= \frac{2\sin\theta\cos\theta - \cos\theta}{1-(1-2\sin^2\theta)-\sin\theta} = \frac{\cos\theta\,(2\sin\theta - 1)}{\sin\theta\,(2\sin\theta-1)} = \cot\theta$$

$$= \tan\left(\frac{1}{2}\pi - \theta\right)$$

$$\Rightarrow \tan(\theta_1 + \theta_2 + \theta_3 + \theta_4) = \tan\left(\frac{1}{2}\pi - \theta\right)$$

$$\Rightarrow \theta_1 + \theta_2 + \theta_3 + \theta_4 = n\pi + \left(\frac{1}{2}\pi - \theta\right)$$

$$\Rightarrow \tan^{-1}\alpha + \tan^{-1}\beta + \tan^{-1}\gamma + \tan^{-1}\delta = n\pi + \left(\frac{1}{2}\pi - \theta\right).$$

Hence proved.

Example 3:

If θ_1, θ_2, θ_3 be three values of θ which satisfy the equation $\tan 2\theta = l \tan(\theta + \alpha)$ and such that no two of them differs by a multiple of π, show that $\theta_1 + \theta_2 + \theta_3 + \alpha$ is a multiple of π.

Solution :

The given equation is $\tan 2\theta = \lambda \tan(\theta + \alpha)$

$$\Rightarrow \qquad \frac{2\tan\theta}{1-\tan^2\theta} = \lambda\left(\frac{\tan\theta + \tan\alpha}{1-\tan\theta\tan\alpha}\right)$$

Cross multiplying, we have

$2\tan\theta\,(1 - \tan\theta\tan\alpha) = \lambda\,(1-\tan^2\theta)(\tan\theta + \tan\alpha)$

$\Rightarrow 2\tan\theta - 2\tan\alpha\tan^2\theta = \lambda\,(\tan\theta + \tan\alpha - \tan^3\theta - \tan\alpha\tan^2\theta)$

$\Rightarrow \lambda\tan^3\theta - (2-\lambda)\tan^2\alpha\tan\theta + (2-\lambda)\tan\theta - \lambda\tan\alpha = 0,$

which is a cubic equation in tan θ. Its roots are given as $\tan\theta_1$, $\tan\theta_2$ and $\tan\theta_3$. Therefore we have

$$\tan\theta_1 + \tan\theta_2 + \tan\theta_3 = -\frac{[(2-\lambda)\tan\alpha]}{\lambda} = \frac{(2-\lambda)\tan\alpha}{\lambda}$$

$\tan\theta_1 \tan\theta_2 + \tan\theta_1 \tan\theta_3 + \tan\theta_2 \tan\theta_3 = (2 - l)/l$ and

$\tan\theta_1 \tan\theta_2 \tan\theta_3 = -[-\lambda \tan\alpha]/i = \tan\alpha$

Also $\tan(\theta_1 + \theta_2 + \theta_3)$

$$= \frac{"s_1 - s_3"}{1 - s_2} = \frac{(\Sigma \tan\theta_1) - (\Sigma \tan\theta_1 \tan\theta_2 \tan\theta_3)}{1 - (\Sigma \tan\theta_1 \tan\theta_2)}$$

$$= \frac{\left[\frac{(2-\lambda)\tan\alpha}{\lambda}\right] - \tan\alpha}{1 - \frac{(2-\lambda)}{\lambda}}$$

$$= \frac{2\tan\alpha - \lambda\tan\alpha - \lambda\tan\alpha}{\lambda - 2 + \lambda} = \frac{(\tan\alpha)(2 - 2\lambda)}{-(2 - 2\lambda)} = -\tan\alpha$$

$\Rightarrow \tan(\theta_1 + \theta_2 + \theta_3) = \tan(-\alpha)$

$\Rightarrow \quad \theta_1 + \theta_2 + \theta_3 = n\pi + (-\alpha) = n\pi - \alpha$

$\Rightarrow \quad \theta_1 + \theta_2 + \theta_3 + \alpha = n\pi$. Where n is zero or any integer.

= a multiple of p. **Hence proved.**

Example 4:

Prove that the equation $\cos 2\theta + a\cos\theta + b\sin\theta + c = 0$ has in general, four roots, and that the sum of the four roots is a multiple of 2π.

Solution :

We know $\sin\theta = \dfrac{2\tan\frac{1}{2}\theta}{1 + \tan^2\frac{1}{2}\theta} = \dfrac{2t}{1 + t^2}$, where $t = \tan\frac{1}{2}\theta$

$$\cos\theta = \frac{1 - \tan^2\frac{1}{2}\theta}{1 + \tan^2\frac{1}{2}\theta} = \frac{1 - t^2}{1 + t^2}$$

And $\cos 2\theta = 1 - 2\sin^2\theta = 1 - 2\,[2t/(1 + t^2)]^2$

$$= \frac{(1 + t^2)^2 - 8t^2}{(1 + t^2)^2} = \frac{1 - 6t^2 + t^4}{(1 + t^2)^2}.$$

Substituting these values in the given equation, we get

$$\frac{1 - 6t^2 + t^4}{(1 + t^2)^2} + a\left(\frac{1 - t^2}{1 + t^2}\right) + b\left(\frac{2t}{1 + t^2}\right) + c = 0$$

$\Rightarrow\ (1-6t^2+t^4)+a(1-t^2)(1+t^2)+2bt(1+t^2)+c(1+t^2)^2=0$

$\Rightarrow\ (1-a+c)t^4+2bt^3+2(c-3)t^2+2bt+(1+a+c)=0.$

The equation is of fourth degree in t or tan $\frac{1}{2}\theta$, hence it has four roots. If t_1, t_2, t_3 and t_4 or tan $\frac{1}{2}\theta_1$, tan $\frac{1}{2}\theta_2$, tan $\frac{1}{2}\theta_3$ and tan $\frac{1}{2}\theta_4$ be its four roots, then

$$\Sigma \tan\frac{1}{2}\theta_1=\frac{-2b}{1-a+c};\ S\tan\frac{1}{2}\theta_1\tan\frac{1}{2}\theta_2=\frac{2(c-3)}{1-a+c};$$

$$\Sigma \tan\frac{1}{2}\theta_1\tan\frac{1}{2}\theta_2\tan\frac{1}{2}\theta_3=\frac{-2b}{1-a+c}$$

$$\text{and } \tan\frac{1}{2}\theta_1\tan\frac{1}{2}\theta_2\tan\frac{1}{2}\theta_3\tan\frac{1}{2}\theta_4=\frac{1+a+c}{1-a+c}$$

$$\text{Also } \tan\left(\frac{1}{2}\theta_1+\frac{1}{2}\theta_2+\frac{1}{2}\theta_3+\frac{1}{2}\theta_4\right)=\frac{"s_1-s_3"}{1-s_2+s_4}$$

$$=\frac{\left(\Sigma\tan\frac{1}{2}\theta\right)-\left(\Sigma\tan\frac{1}{2}\theta_1\tan\frac{1}{2}\theta_2\tan\frac{1}{2}\theta_3\right)}{1-\left(\Sigma\tan\frac{1}{2}\theta_1\tan\frac{1}{2}\theta_2\right)+\left(\tan\frac{1}{2}\theta_1\tan\frac{1}{2}\theta_2\tan\frac{1}{2}\theta_3\tan\frac{1}{2}\theta_4\right)}$$

$$=\frac{\left(\frac{-2b}{1-a+c}\right)-\left(\frac{-2b}{1-a+c}\right)}{1-\frac{2(c-5)}{1-a+c}+\frac{1+a+c}{1-a+c}}=0$$

$\Rightarrow\ \frac{1}{2}\theta_1+\frac{1}{2}\theta_2+\frac{1}{2}\theta_3+\frac{1}{2}\theta_4=n\pi \Rightarrow \theta_1+\theta_2+\theta_3+\theta_4=2n\pi$

$\Rightarrow\ \theta_1+\theta_2+\theta_3+\theta_4=$ a multiple of 2π.

Hence proved.

Example 5:

Prove that the equation $a^2\cos^2\theta+b^2\sin^2\theta+2g\,a\cos\theta+2fb\sin\theta+c=0$ *has four roots and that the sum of the values of* θ *which satisfy it is an even multiple of* π *radians.*

Or

If θ_1, θ_2, θ_3 *and* θ_4 *be the eccentric angles of the points of intersection of the ellipse* $x=a\cos\theta$, $y=b\sin\theta$ *and the circle* $x^2+y^2+2gx+2fy+c=0$ *show that* $\theta_1+\theta_2+\theta_3+\theta_4=2n\pi$.

Solution :

We know $\sin\theta = \dfrac{2\tan\frac{1}{2}\theta}{1+\tan^2\frac{1}{2}\theta}$ and $\cos\theta = \dfrac{1-\tan^2\frac{1}{2}\theta}{1+\tan^2\frac{1}{2}\theta}$

Putting $\tan\frac{1}{2}\theta = t$, we have

$\sin\theta = 2t/(1+t^2)$ and $\cos\theta = (1-t^2)/(1+t^2)$ ...(i)

Substituting these values of $\sin\theta$ and $\cos\theta$ in the given equation, it becomes

$$a^2\left(\frac{1-t^2}{1+t^2}\right)^2 + b^2\left(\frac{2t}{1+t^2}\right)^2 + 2ga\left(\frac{1-t^2}{1+t^2}\right) - 2fb\left(\frac{2t}{1+t^2}\right) + c = 0$$

$$\Rightarrow a^2(1-t^2)^2 + b^2(4t^2) + 2ga(1-t^2)(1+t^2) + 2fb(2t)(1+t^2) + c(1+t^2)^2 = 0$$

$$\Rightarrow a^2(1-2t^2+t^4) + b^2(4t^2) + 2ga(1-t^4) + 4fbt(1+t^2) + c(1+2t^2+t^4) = 0$$

$$\Rightarrow t^4(a^2-2ga+c) + 4fbt^3 + t^2(4b^2-2a^2+2c) + 4fbt + (a^2+2ga+c) = 0$$

The equation is of fourth degree in t or $\tan\frac{1}{2}\theta$, hence it has four roots.

If t_1, t_2, t_3 and t_4 or $\tan\frac{1}{2}\theta_1$, $\tan\frac{1}{2}\theta_2$, $\tan\frac{1}{2}\theta_3$, $\tan\frac{1}{2}\theta_4$ be its four roots then

$$\Sigma\tan\frac{1}{2}\theta_1 = \frac{-4fb}{a^2+2ga+c^2} \quad \Sigma\tan\frac{1}{2}\theta_1\tan\frac{1}{2}\theta_2 = \frac{4b^2-2a^2+2c}{a^2-2ga+c},$$

$$\Sigma\tan\frac{1}{2}\theta_1\tan\frac{1}{2}\theta_2\tan\frac{1}{2}\theta_3 = \frac{-4fb}{a^2-2ga+c}$$

and $$\tan\frac{1}{2}\theta_1\tan\frac{1}{2}\theta_2\tan\frac{1}{2}\theta_3\tan\frac{1}{2}\theta_4 = \frac{a^2+2ag+c}{a^2-2ga+c}$$

Also $$\left(\tan\frac{1}{2}\theta_1 + \frac{1}{2}\theta_2 + \frac{1}{2}\theta_3 + \frac{1}{2}\theta_4\right) = \frac{"s_1-s_3"}{s_2+s_4}$$

$$= \frac{\left(\Sigma\tan\frac{1}{2}\theta_1\right) - \left(\Sigma\tan\frac{1}{2}\theta_1\tan\frac{1}{2}\theta_2\tan\frac{1}{2}\theta_3\right)}{1-\left(\Sigma\tan\frac{1}{2}\theta_1\tan\frac{1}{2}\theta_2\right) + \left(\Sigma\tan\frac{1}{2}\theta_1\tan\frac{1}{2}\theta_2\tan\frac{1}{2}\theta_3\tan\frac{1}{2}\theta_4\right)}$$

$$= \frac{\left(\frac{-4fb}{a^2-2ga+c}\right)-\left(\frac{-4fb}{a^2-2ga+c}\right)}{1-\left(\frac{4b^2-a^2+2c}{a^2-2ga+c}\right)+\left(\frac{a^2+2ga+c}{a^2-2ga+c}\right)} = \frac{0}{\text{some thing}} = 0$$

$$\Rightarrow \tan\left(\frac{1}{2}\theta_1+\frac{1}{2}\theta_2+\frac{1}{2}\theta_3+\frac{1}{2}\theta_4\right)=0$$

$$\Rightarrow \frac{1}{2}\theta_1+\frac{1}{2}\theta_2+\frac{1}{2}\theta_3+\frac{1}{2}\theta_4 = n\pi$$

$\Rightarrow \theta_1+\theta_2+\theta_3+\theta_4 = n\pi$ = even multiple of π. **Hence proved.**

Example 6:

Prove that the equation as sec θ – bk cosec θ = $a^2 - b^2$, has four roots and that sum of the values of θ, which satisfy it, is equal to an odd multiple of π radians.

Solution :

We know $\sin\theta = \dfrac{2\tan\frac{1}{2}\theta}{1+\tan^2\frac{1}{2}\theta}$ and $\cos\theta = \dfrac{1-\tan^2\frac{1}{2}\theta}{1+\tan^2\frac{1}{2}\theta}$

$\therefore$ $\operatorname{cosec}\theta = \dfrac{1+\tan^2\frac{1}{2}\theta}{2\tan\frac{1}{2}\theta}$ and $\sec\theta = \dfrac{1+\tan^2\frac{1}{2}\theta}{1-\tan^2\frac{1}{2}\theta}$

$\Rightarrow$ $\operatorname{cosec}\theta = \dfrac{1+t^2}{2t}$ and $\sec\theta = \dfrac{1+t^2}{1-t^2}$ where $t = \tan\frac{1}{2}\theta$

Making these substitutions in the given equation we have

$$ah\left(\frac{1+t^2}{1-t^2}\right) - bk\left(\frac{1+t^2}{2t}\right) = a^2 - b^2$$

$\Rightarrow$ $2ah(1+t^2)t - bk(1+t^2)(1-t^2) = (a^2-b^2)2t(1-t^2)$

$\Rightarrow$ $2ah(t+t^3) - tk(1-t^4) = (a^2-b^2)(2t-2t^3)$

$\Rightarrow$ $bk\,t^4 + (2ah+2a^2-2b^2)t^3 + (0)t^2 + (2ah-2a^2+2b^2)t - bk = 0$

This equation is of fourth degree in t_2 hence it has four roots.

Let its roots be t_1, t_2, t_3 and t_4 or $\tan\frac{1}{2}\theta_1$, $\tan\frac{1}{2}\theta_2$, $\tan\frac{1}{2}\theta_3$ and $\tan\frac{1}{2}\theta_4$

Then s_1 = sum of roots taken once

$$= \Sigma\tan\frac{\theta_1}{2} = \frac{-(2ah+2a^2-2b^2)}{bk}$$

s_2 = sum taken two at a time = $\Sigma \tan\frac{\theta_1}{2}\tan\frac{\theta_2}{2} = \frac{0}{bk} = 0$;

s_3 = sum taken three at a time = $\Sigma \tan\frac{1}{2}\theta_1 \tan\frac{1}{2}\theta_2 \tan\frac{1}{2}\theta_3$

$= -(2ah\ 2a^2 + 2b^2)/bk$;

$$s_4 = \Sigma \tan\frac{1}{2}\theta_1 \tan\frac{1}{2}\theta_2 \tan\frac{1}{2}\theta_3 \tan\frac{1}{2}\theta_4 = -bk/bk = -1$$

$$\text{Also } \tan\left(\frac{1}{2}\theta_1 + \frac{1}{2}\theta_2 + \frac{1}{2}\theta_3 + \frac{1}{2}\theta_4\right) = \frac{\text{"}s_1 + s_3\text{"}}{1 - s_2 + s_4}$$

$$= \frac{\frac{-(2ah + 2a^2 - 2b^2)}{bk} + \frac{(2ah - 2a^2 + 2h^2)}{bk}}{1 - 0 + (-1)} = \frac{4(b^2 - a^2)}{0}$$

$$= \infty \tan\frac{1}{2}\pi$$

$\Rightarrow \frac{1}{2}\theta_1 + \frac{1}{2}\theta_2 + \frac{1}{2}\theta_3 + \frac{1}{2}\theta_4 = n\pi + \frac{1}{2}\pi$, where n is any integer

$\Rightarrow \theta_1 + \theta_2 + \theta_3 + \theta_4 = 2n\pi + \pi = (2n + 1)\pi$

= odd multiple of p, as (2n + 1) is an

odd integer for all integral values of n. **Hence proved.**

SOLVED PROBLEMS

Example 1:

If $\frac{\sin\theta}{\theta} = \frac{2165}{2166}$, *prove that* θ *is the number of radians in* 3° *nearly.*

Solution :

$\frac{2165}{2166}$ is nearly equal to. 1.

$\therefore \frac{\sin\theta}{\theta}$ is also nearly equal to 1 and so θ is very small.

$$\text{Now } \frac{\sin\theta}{\theta} = \frac{2165}{2166} \Rightarrow \frac{\theta - (\theta^3/3!) + (\theta^5/5!) - \ldots}{\theta} = \frac{2165}{2166}$$

$\Rightarrow 1 - \frac{\theta^2}{3!} = \frac{2165}{2166}$, neglecting higher powers of θ as it is very small.

$$\Rightarrow \frac{\theta^2}{3!} = 1 - \frac{2165}{2166} = \frac{1}{2166} \Rightarrow \theta^2 = \frac{6}{2166} = \frac{1}{361}$$

$\Rightarrow\ \theta = 1/19$ radians

$= \dfrac{1}{19}\times\dfrac{180}{\pi}$ degrees, $\quad \because\ \pi$ radians = 180°

$= \dfrac{180\times 7}{19\times 22}$ degrees = 3° nearly. **Hence proved.**

Example 2:

If $\dfrac{\sin\theta}{\theta} = \dfrac{5765}{5766}$, *find* θ.

Solution :

$\therefore\ \dfrac{5765}{5766}$ is nearly equal to 1.

$\therefore\ \dfrac{\sin\theta}{\theta}$ is also nearly equal to 1 and so θ is very small.

Now $\dfrac{\sin\theta}{\theta} = \dfrac{5765}{5766} \Rightarrow \dfrac{\theta-(\theta^3/3!)+(\theta^5/5!)-\ldots}{\theta} = \dfrac{5765}{5766}$

$\Rightarrow\ 1-\dfrac{\theta^2}{3!} = \dfrac{5765}{5766}$, neglecting higher powers of small θ.

$\Rightarrow\ \dfrac{\theta^2}{6} = 1-\dfrac{5765}{5766} = \dfrac{1}{5766} \Rightarrow \theta^2 = \dfrac{1}{961}$

$\Rightarrow\ \theta = \dfrac{1}{31}$ radians $= \dfrac{1}{31}\times\dfrac{180}{\pi}$ degrees, $\therefore\ \pi$ radians = 180°

$= \dfrac{180\times 7}{31\times 22}$ degrees = 1.84° degrees = 2° nearly. **Hence Proved**

Example 3:

If cos θ = θ, prove that θ is equal to 0.7 radian nearly.

Solution :

Given $\cos\theta = \theta \Rightarrow 1-\dfrac{\theta^2}{2!}+\dfrac{\theta^4}{4!}-\ldots = \theta$

$\Rightarrow\ 1-\theta^2/2! = 1$, neglecting powers of θ higher than the second.

$\Rightarrow\ 2-\theta^2 = 2\theta \Rightarrow \theta^2+2\theta-2 = 0$

$\Rightarrow\ \theta = \dfrac{1}{2}\left[-2+\sqrt{(4+8)}\right] = -1+\sqrt{3}$ = 732 radians,

= 0.7 radians nearly. **Hence proved.**

Example 4:

If $\tan \theta = \frac{1}{15}$, *find an approximate value for* θ.

Solution :

If $\tan\theta = \frac{1}{15}$, then $\theta + \frac{\theta^3}{3} + ... = \frac{1}{15}$

$\Rightarrow \quad \theta = \frac{1}{15}$, neglecting powers of θ higher than the first

$$\Rightarrow \quad \theta = \frac{1}{15} \text{ radians } = \frac{1}{15} \times \frac{180}{\pi} \text{ degrees } = \frac{1}{15} \times \frac{180 \times 7}{22}$$

$= 42/11 = 3^\circ\, 49'$ nearly. **Ans.**

Example 5:

If $\sin\left(\frac{1}{6}\pi + \theta\right) = 0.51$, *prove that* θ *is equal to 40' nearly.*

Solution :

$\because$ 0.51 is nearly equal to $\frac{1}{2}$ *i.e.*, $\sin \frac{1}{6}\pi$

$\therefore$ θ must be small.

Now the given equation is $\sin\left(\frac{1}{6}\pi + \theta\right) = 0.51$

$$\Rightarrow \quad \sin \frac{1}{6}\pi \cos\theta + \cos \frac{1}{6}\pi \sin\theta = 0.51$$

$$\Rightarrow \quad \frac{1}{2} \cos\theta + \left(\sqrt{3/2}\right) \sin\theta = 0.51$$

$$\Rightarrow \quad \frac{1}{2}\left(1 - \frac{\theta^2}{2!} + ...\right) + \frac{\sqrt{3}}{2}\left(\theta - \frac{\theta^3}{3!} + ...\right) = 0.51$$

$\frac{1}{2}\left(1 - \frac{1}{2}\theta^2\right) + \frac{1}{2}\sqrt{3}\,(\theta) = 0.51$, neglecting powers of θ higher than the second.

$$\Rightarrow \quad -\frac{1}{4}\theta^2 + \frac{1}{2}\sqrt{3q} = .51 - \frac{1}{2} = 0.51 - 0.50 = 0.01 = \frac{1}{100}$$

$$\Rightarrow \quad 25\theta^2 - 50\sqrt{3\theta} + 1 = 0$$

$$\Rightarrow \ \theta = 0.012 \text{ radian} = \frac{12}{1000} \times \frac{180 \times 7}{22} \times 60 \text{ min} = 40' \text{ nearly.}$$

Example 6:

If $\cos\left(\frac{1}{3}\pi + \theta\right) = 0.49$, *prove that θ is equal to 40′ nearly.*

Solution :

0.49 is nearly equal to $\frac{1}{2}$ *i.e.*, $\cos \frac{1}{2}\pi$

∴ θ must be very small.

Now the given equation is $\cos\left(\frac{1}{3}\pi + \theta\right) = 0.49$

$$\Rightarrow \ \cos\left(\frac{1}{2}\pi\right)\cos\theta + \sin\left(\frac{1}{2}\pi\right)\sin\theta = 0.49$$

$$\Rightarrow \ \frac{1}{2}\cos\theta - \left(\sqrt{3/2}\right)\sin\theta = 0.49$$

$$\Rightarrow \ \frac{1}{2}\left(1-\frac{\theta^2}{2!}+\ldots\right) - \frac{\sqrt{3}}{2}\left(\theta-\frac{\theta^3}{3!}+\ldots\right) = 0.49$$

$$\Rightarrow \ \frac{1}{2}\left(1-\frac{\theta^2}{2!}\right) - \frac{\sqrt{3}}{2}(\theta) = 0.49$$, neglecting powers of θ higher than

the second.

$$\frac{1}{2}-\frac{\theta^2}{4}-\frac{\theta\sqrt{3}}{2} = C.49 = \frac{49}{100} \Rightarrow 25\theta^2 + 50\sqrt{3}\theta - 1 = 0$$

$$\Rightarrow \ \theta = 0.12 \text{ radian nearly} = \frac{12}{1000} \times \frac{180}{22} \times 7 \times 60 \text{ min.}$$

$$= 40 \text{ minutes nearly.}$$ **Hence proved.**

Example 7:

Calculate the value of sin 3° correct to three significant figures.

Solution :

$$3^\circ = 3 \times \frac{\pi}{180} \text{ radian} = \frac{3 \times 22}{7 \times 180} = \frac{11}{210} \text{ radian}$$

Now we know that $\sin\alpha = \alpha - \frac{\alpha^3}{3!} + \frac{\alpha^5}{5!} + \ldots$

$$\therefore \sin 3^\circ = \sin\frac{11}{210} = \frac{11}{210} - \frac{1}{3\,i}\left(\frac{11}{210}\right)^3 + \ldots$$

$$= .0523 - \frac{1}{6}(.000143) + \ldots$$

$$= .0523 - .000024 + \ldots = .0523.$$ **Ans.**

Example 8:

Prove that

$$\frac{1}{6}\sin^2\theta = \frac{\theta^3}{3!} - \left(1+3^2\right)\frac{\theta^5}{5!} + \left(1+3^2+3^4\right)\frac{\theta^7}{7} - \ldots$$

Solution :

We know that $\sin 3\theta = 3\sin\theta - 4\sin^3\theta$

$$\Rightarrow \quad 4\sin^3\theta = 3\sin\theta - \sin 3\theta$$

$$\Rightarrow \quad \sin^3\theta = \frac{1}{4}[3\sin\theta - \sin 3\theta]$$

$$\Rightarrow \quad \sin^3\theta = \frac{1}{4}\left[3\left\{\theta - \frac{\theta^3}{3!} + \frac{\theta^5}{5!} - \frac{\theta^7}{7!} + \ldots\right\} - \left\{(3\theta) - \frac{(3\theta)^3}{3!} + \frac{(3\theta)^5}{5!} - \frac{(3\theta)^7}{7!} + \ldots\right\}\right]$$

$$= \frac{1}{4}\left[(3\theta - 3\theta) - \frac{\theta^3}{3!}(3-3)^3 + \frac{\theta^5}{5!}(3-3^5) - \frac{\theta^7}{7!}(3-3^7) + \ldots\right]$$

$$= \frac{1}{4}\left[\frac{\theta^3}{3!}(3-3)^3 - \frac{\theta^5}{5!}(3^5-3) + \frac{\theta^7}{7!}(3^7-3) - \ldots\right]$$

$$= \frac{3}{4}\left[\frac{\theta^3}{3!}(3-1) - \frac{\theta^5}{5!}(3^4-1) + \frac{\theta^7}{7!}(3^6-1) - \ldots\right]. \qquad \ldots(i)$$

taking 3 common from each term.

$$= \frac{3}{4}(3^2-1)\left[\frac{\theta^3}{3!} - \frac{\theta^5}{5!}(3^2+1) + \frac{\theta^7}{7!}(3^4+3^2+1) - \ldots\right]$$

$$\sin^3 = \frac{3}{4}\times 8 = \left[\frac{\theta^3}{3!} - \frac{\theta^5}{5!}\left(3^2+1\right) + \frac{\theta^7}{7!}\left(3^4+3^2+1\right) - \ldots\right]$$

$$\Rightarrow \frac{1}{6}\sin^3\theta = \frac{\theta^3}{3!} - \frac{\theta^5}{5!}\left(3^2+1\right) + \frac{\theta^7}{7!}\left(3^4+3^2+1\right) - \ldots$$ **Hence proved.**

Example 9:

Prove that

$$\sin^2\theta\cos\theta = \theta^2 - \frac{5}{6}\theta^4 + \ldots + (-1)^{n+1}\frac{3^{2n}-1}{4}\cdot\frac{\theta^{2n}}{2n!} + \ldots$$

Solution :

$$\sin^2\theta\cos\theta = \sin\theta.\sin\theta\cos\theta = \frac{1}{2}\sin\theta.\sin 2\theta$$

$$= \frac{1}{4}(2\sin\theta\sin 2\theta) = \frac{1}{4}(\cos\theta - \cos 3\theta)$$

$$= \frac{1}{4}\left[\left\{1 - \frac{\theta^2}{2!} + \frac{\theta^4}{4!} - \ldots + (-1)^n\frac{\theta^{2n}}{2n!} + \ldots\right\}\right.$$

$$\left. - \left\{1 - \frac{(3\theta)^2}{2!} + \frac{(3\theta)^4}{4!} - \ldots + \frac{(-1)^n(3\theta)^{2n}}{(2n)!} + \ldots\right\}\right]$$

$$= \frac{1}{4}\left[-\frac{\theta^2}{2!}\left(1-3^2\right) + \frac{\theta^4}{4!}\left(1-3^4\right) - \ldots + (-1)^n\frac{\theta^{2n}}{(2n)!}\left(1-3^{2n}\right) + \ldots\right]$$

$$= \frac{1}{4}\left[4\theta^2 - \frac{10}{3}\theta^4 + \ldots + \frac{(-1)^n(-1)\left(3^{2n}-1\right)\theta^{2n}}{(2n)!} + \ldots\right]$$

$$= \theta^2 - \frac{5}{6}\theta^4 + \ldots + \frac{(-1)^{n+1}\left(3^{2n}-1\right)\theta^{2n}}{(2n)!} + \ldots$$

Example 10:

Expand (sin θ)/θ in powers of θ.

Solution :

$$\frac{\sin\theta}{\theta} = \frac{1}{\theta}\left[\theta - \frac{\theta^2}{3!} + \frac{\theta^5}{5!} - \ldots \text{ ad. anf.}\right]$$

$$= 1 - \frac{\theta^2}{3!} + \frac{\theta^4}{5!} - \frac{\theta^6}{7!} - \ldots \text{ ad. inf.}$$ **Ans.**

Example 11:

If θ be small, prove that $\theta \cot \theta = 1 - \dfrac{\theta^2}{3} - \dfrac{\theta^4}{45}$ *approx.*

Solution :

$$\theta \cot \theta = \frac{\theta}{\tan \theta} = \frac{\theta}{\theta + \frac{1}{3}\theta^3 + \frac{2}{15}\theta^5}\text{, neglecting higher powers of } \theta.$$

$$\Rightarrow \quad \theta \cot \theta = \frac{\theta}{1 + \frac{1}{3}\theta^2 + \frac{2}{15}\theta^4} = \left[1 + \frac{1}{3}\theta^2 + \frac{2}{15}\theta^4\right]^{-1}$$

$$= 1 - \left(\frac{1}{3}\theta^2 + \frac{2}{15}\theta^4\right) + \left(\frac{1}{2}\theta^2 + \frac{2}{15}\theta^4\right)^2 - \ldots$$

$$= 1 - \frac{1}{3}\theta^2 - \frac{2}{15}\theta^4 + \frac{1}{9}\theta^4\text{, neglecting higher powers of } \theta$$

$$= 1 - \frac{1}{3}\theta^2 - \left(\frac{2}{15} - \frac{1}{9}\right)\theta^4 = 1 - \frac{1}{3}\theta^2 - \frac{1}{45}\theta^4 \text{ approx.}$$

Example 12:

Determine a, b, c, if

$$\lim_{\theta \to 0} \frac{\theta\left(a + b\cos\theta\right) - c\sin\theta}{\theta^5} = 1.$$

Solution :

$$\frac{\theta\left(a + b\cos\theta\right) - c\sin\theta}{\theta^5} = 1$$

$$= \frac{1}{\theta^5}\left[\theta\left\{a + b\left(\theta - \frac{\theta^2}{2!} + \frac{\theta^4}{4!} - \frac{\theta^6}{6!} + \ldots\right)\right\} - c\left\{\theta - \frac{\theta^3}{3!} + \frac{\theta^5}{5!} - \frac{\theta^7}{7!} + \ldots\right\}\right],$$

expanding cos θ and sin θ

$$= \frac{1}{\theta^5}\left[\left(a + b + c\right)\theta + \left(\frac{c}{3!} - \frac{b}{2!}\right)\theta^3 + \left(\frac{b}{4!} - \frac{c}{5!}\right)\theta^5 + \left(\frac{c}{7!} - \frac{b}{6!}\right)\theta^7 + \ldots\right]$$

$$= \frac{(a+b+c)}{\theta^4} + \left(\frac{c}{3!} - \frac{b}{2!}\right)\frac{1}{\theta^2} + \left(\frac{b}{4!} - \frac{c}{5!}\right) + \left(\frac{c}{7!} - \frac{b}{6!}\right)\theta^2 + \ldots$$

$\therefore$ If $\lim\limits_{\theta \to 0} \frac{(a + b\cos\theta) - c\sin\theta}{\theta^5} = 1$, then we must have

$$a + b - c = 0,\ \frac{c}{3!} - \frac{b}{2!} = 0 \text{ and } \frac{b}{4!} - \frac{c}{5!} = 1$$

$\Rightarrow$ $a + b - c = 0;\ c - 3b = 0,\ 5b - c = 120$

Solving these we get $a = 120$, $b = 60$, $c = 180$. **Ans.**

Example 13:

Show that $\frac{\pi^2}{2.4} - \frac{\pi^4}{2.4.6.8} + \frac{\pi^6}{2.4.6.8.10.12} - \ldots = 1$.

Solution :

We know $\cos\theta = 1 - \frac{\theta^2}{2!} + \frac{\theta^4}{4!} - \frac{\theta^6}{6!} + \ldots$

Putting $\theta = \frac{1}{2}\pi$ on both sides we have

$$\cos\frac{1}{2}\pi = 1 - \frac{\pi^2}{2^2\,2!} + \frac{\pi^4}{2^4\,4!} - \frac{\pi^6}{2^6\,6!} + \ldots$$

$$\Rightarrow\ 0 = 1 - \frac{\pi^2}{2 \times 4} + \frac{\pi^4}{16 \times 24} - \frac{\pi^6}{64 \times 720} + \ldots$$

$$\Rightarrow\ 0 = 1 - \frac{\pi^2}{2.4} + \frac{\pi^4}{2.4.6.8} - \frac{\pi^6}{2.4.6.8.10.12} + \ldots$$

$$\Rightarrow\ \frac{\pi^2}{2.4} - \frac{\pi^4}{2.4.6.8} + \frac{\pi^6}{2.4.6.8.10.12} - \ldots = 1.$$

Hence proved.

Example 14:

If $x = \frac{2}{1!} - \frac{4}{3!} + \frac{6}{5!} - \frac{8}{7!} + \ldots$ *ad. inf.*

and $y = 1 + \frac{2}{1!} - \frac{2^3}{3!} + \frac{2^5}{5!} - \ldots$ *ad. inf., then show that* $x^2 = y$.

Solution :

We know that $\sin\theta = \theta - \frac{\theta^3}{3!} + \frac{\theta^5}{5!} - \ldots$ ad. inf.

Putting $\theta = 2$ on both sides we have

$$\sin 2 = 2 - \frac{2^3}{3!} + \frac{2^5}{5!} - \dots \text{ ad. inf.}$$

$$\therefore\ y = 1 + \left(\frac{2}{1!} - \frac{2^3}{3!} + \frac{2^5}{5!} - \dots\text{ad. inf.}\right) \Rightarrow y = 1 + \sin 2 \qquad \dots(i)$$

Again $x = \frac{2}{1!} - \frac{4}{3!} + \frac{6}{5!} - \frac{8}{7!} + \dots$ ad. inf.

$$= \frac{2+1}{1!} - \frac{3+1}{3!} + \frac{5+1}{5!} - \frac{7+1}{7!} + \dots \text{ ad. inf.}$$

$$= \left(\frac{1}{1!} - \frac{3}{3!} + \frac{5}{5!} - \frac{7}{7!} + \dots\text{ad. anf.}\right)$$

$$+ \left(\frac{1}{1!} - \frac{1}{3!} - \frac{1}{5!} - \frac{1}{7!} + \dots\text{ad. anf.}\right)$$

$$= \left(1 - \frac{1}{2!} + \frac{1}{4!} - \frac{1}{6!} + \dots\text{ad. anf.}\right)$$

$$+ \left(1 - \frac{1}{3!} + \frac{1}{5!} - \frac{1}{7!} + \dots\text{ad. anf.}\right)$$

$= \cos 1 + \sin 1$ [We can verify by putting $\theta = 1$ in the expansions of $\sin\theta$ and $\cos\theta$]

$\therefore\ x^2 = (\cos 1 + \sin 1)^2 = \cos^2 1 + \sin^2 1 + 2 \sin 1 \cos 1$

$= 1 + \sin 2 = y$, from (i)

$\Rightarrow x^2 = y$ **Hence proved.**

Example 15:

Prove that $\left(x - 2\cos\frac{2\pi}{5}\right)\left(x - 2\cos\frac{4\pi}{5}\right)\left(x - 2\cos\frac{6\pi}{5}\right)$

$$\times \left(x - 2\cos\frac{8\pi}{5}\right) = x^4 + 2x^3 - x^2 - 2x + 1$$

Solution :

$$\left(x - 2\cos\frac{2\pi}{5}\right)\left(x - 2\cos\frac{4\pi}{5}\right)\left(x - 2\cos\frac{6\pi}{5}\right)\left(x - 2\cos\frac{8\pi}{5}\right)$$

$$= \left(x - 2\cos\frac{2\pi}{5}\right)\left(x - 2\cos\frac{4\pi}{5}\right)\left(x - 2\cos\frac{4\pi}{5}\right)\left(x - 2\cos\frac{2\pi}{5}\right),$$

$$\because \quad \cos\frac{6\pi}{5} = \cos\left(2\pi - \frac{4\pi}{5}\right) = \cos\frac{4\pi}{5} \text{ and } \cos\frac{8}{5}\pi = \cos\frac{2}{5}\pi$$

$$= \left[\left(x - 2\cos\frac{2\pi}{5}\right)\left(x - 2\cos\frac{4\pi}{5}\right)\right]^2$$

$$= \left[x^2 - 2x\left(\cos\frac{2\pi}{5} + \cos\frac{4\pi}{5}\right) + 4\cos\frac{2\pi}{5}\cos\frac{4\pi}{5}\right]^2$$

$$= [x^2 - 2x(\cos 72^\circ + \cos 144)^\circ + 4\cos 72^\circ \cos 144^\circ]^2$$

$$= [x^2 - 2x(\sin 18^\circ - \cos 36^\circ) + 4(\sin 18^\circ)(-\cos 36^\circ)]^2$$

... $\because$ $\cos 72^\circ = \cos(90^\circ - 18^\circ) = \sin 18^\circ$

and $\cos 144^\circ = \cos(\pi - 36^\circ) = -\cos 36^\circ$

$$= \left[x^2 - 2x\left(\frac{\sqrt{5}-1}{4} - \frac{\sqrt{5}+1}{4}\right) + 4\left(\frac{\sqrt{5}-1}{4}\right)\left(-\frac{\sqrt{5}+1}{4}\right)\right]^2$$

$$= (x^2 + x - 1)^2 = x^4 + x^2 + 1 + 2x^3 - 2x^3 - 2x^2 - 2x = x^4 + 2x^3 - x^2 - 2x + 1$$

Example 16:

(a) Prove that the roots of the equation

$8x^3 - 4x^2 - 4x + 1 = 0$ are $\cos\left(\frac{1}{7}\pi\right)$, $\cos\left(\frac{3}{7}\pi\right)$ and $\cos\left(\frac{5}{7}\pi\right)$

Hence obtain the equations whose roots are

(b) $\sec^2\left(\frac{1}{7}\pi\right)$, $\sec^2\left(\frac{3}{7}\pi\right)$, $\sec^2\left(\frac{5}{7}\pi\right)$.

(c) $\tan^2\left(\frac{1}{7}\pi\right)$, $\tan^2\left(\frac{3}{7}\pi\right)$, $\tan^2\left(\frac{5}{7}\pi\right)$.

(d) evaluate $\sec\left(\frac{1}{7}\pi\right) + \sec\left(\frac{3}{7}\pi\right) + \sec\left(\frac{5}{7}\pi\right)$.

and evaluate $\sec\left(\frac{1}{7}\pi\right).\cos\left(\frac{3}{7}\pi\right).\cos\left(\frac{5}{7}\pi\right)$.

Solution :

(a) Let θ have one of the following seven values

$$\frac{\pi}{7}, \frac{3\pi}{7}, \frac{5\pi}{7}, \frac{7\pi}{7}, \frac{11\pi}{7} \text{ and } \frac{13\pi}{7} \qquad \textbf{(Note)} \qquad ...(i)$$

i.e., $\theta = \frac{1}{7}(2n+1)\pi$, where n = 0, 1, 2, ..., 6

Then $\quad 7\theta = (2n+1)\pi \Rightarrow 4\theta = (2n+1)\pi - 3\theta$ **(Note)**

$\Rightarrow \cos 4\theta = \cos[(2n+1)\pi - 3\theta] = \pi - 3\theta] = -\cos 3\theta$ ...(ii) **(Note)**

Now $\cos\theta = 4\cos^3\theta - 3\cos\theta$ and

$\cos 4\theta = 2\cos^2 2\theta - 1 = 2(2\cos^2\theta - 1)^2 - 1$

$= 2(4\cos^4\theta - 4\cos^2\theta + 1) - 1 = 8\cos^4\theta - 8\cos^2\theta + 1$

Substituting these values of $\cos 4\theta$ and $\cos 3\theta$ in (ii) we get

$$8\cos^4\theta - 8\cos^2\theta + 1 = -(4\cos^3\theta - 3\cos\theta)$$

$$\Rightarrow \quad 8x^4 + 4x^3 - 8x^2 - 3x + 1 = 0, \text{ where } x = \cos\theta$$

$$\Rightarrow \quad (x+1)(8x^3 - 4x^2 - 4x + 1) = 0 \qquad \text{...(iii)}$$

Now $x = \cos\theta$ and θ can take one of the seven values given in (i).

Also $\quad \cos\frac{13\pi}{7} = \cos\left(2\pi - \frac{\pi}{7}\right) = \cos\frac{\pi}{7}$;

$\cos\frac{11\pi}{7} = \cos\left(2\pi - \frac{3\pi}{7}\right) = \cos\frac{3\pi}{7}$;

and $\cos\frac{9\pi}{7} = \cos\left(2\pi - \frac{5\pi}{7}\right) = \cos\frac{5\pi}{7}$.

Also $\cos\frac{1}{7}(7\pi) = \cos\pi = -1$ *i.e.,* the factor $(x+1)$ in (iii) corresponds to the value $\frac{1}{7}(7\pi)$ of θ.

Hence the equation whose roots are $\cos\frac{1}{7}\pi$, $\cos\frac{3}{7}\pi$, $\cos\frac{5}{7}\pi$ is

$$8x^3 - 4x^2 - 4x + 1 = 0. \qquad \text{...(iv)}$$

(b) In above part we have proved that the roots of the equation $8x^3 - 4x^2 - 4x + 1 = 0$ are $\cos\left(\frac{1}{7}\pi\right)$, $\cos\left(\frac{3}{7}\pi\right)$ and $\cos\left(\frac{5}{7}\pi\right)$

$\therefore \sec^2\frac{\pi}{7} = 1/\cos^2\frac{\pi}{7} \quad \therefore$ Put $y = \frac{1}{x^2}$

$\Rightarrow x = 1/\sqrt{y}$ in the above equation

Then $\sec^2\frac{1}{7}\pi$, $\sec^2\left(\frac{3}{7}\pi\right)$ and $\sec^2\left(\frac{5}{7}\pi\right)$ are the roots of the equation

$$8\left(1/\sqrt{y}\right)^2 - 4\left(1/\sqrt{y}\right)^2 - 4\left(1/\sqrt{y}\right) + 1 = 0$$

$$\Rightarrow \left(8/y\sqrt{y}\right) - (4/y) - \left(4/\sqrt{y}\right)^2 + 1 = 0 \Rightarrow 8 - 4\sqrt{y} - 4y + y\sqrt{y} = 0$$

$$\Rightarrow 8 - 4y = 4\sqrt{y} - y\sqrt{y} = \sqrt{y}\,(4 - y)$$

$$\Rightarrow (8 - 4y)^2 = y\,(4 - y)^2 \text{ Þ } y^3 - 24y^2 + 80y - 64 = 0.$$ **Ans.**

The sum of its roots = 24

i.e., $\sec^2\left(\frac{1}{7}\pi\right) + \sec^2\left(\frac{3}{7}\pi\right) + \sec^2\left(\frac{5}{7}\pi\right) = 24.$

(c) In part (b) above we have proved the equation

$$y^3 - 2y^2 + 80y - 64 = 0$$

has its roots as $\sec^2 \frac{1}{7}\pi$, $\sec^2 \frac{3}{7}\pi$, and $\sec^2 \frac{5}{7}\pi$.

Now we are to obtain the equation whose roots are

$$\tan^2 \frac{1}{7}\pi,\ \tan^2 \frac{3}{7}\pi \text{ and } \tan^2 \frac{5}{7}\pi$$

Now $\tan^2 \frac{1}{7}\pi = \left(\sec^2 \frac{1}{7}\pi\right) - 1$

$\therefore$ Put $z = y - 1 \Rightarrow y = z + 1$ in equation (iii).

The we have required equation as

$$(z + 1)^3 - 24\,(z + 1)^2 + 80\,(z + 1) - 64 = 0$$

$$\Rightarrow (z^2 + 3z^2 + 3z + 1) - 24\,(z^2 + 2z + 1) + 80\,(z + 1) - 64 = 0$$

$$\Rightarrow z^3 - + z^2 + 35z - 7 = 0$$ **Ans.**

(d) We have proved in part (a) result (ii) that the roots of the[2] equation

$8x^3 - 4x^2 - 4x + 1 = 0$ are $\cos \frac{1}{7}\pi$, $\cos \frac{3}{7}\pi$, $\cos \frac{5}{7}\pi$

$$\therefore \quad \cos \frac{1}{7}\pi + \cos \frac{3}{7}\pi + \cos \frac{5}{7}\pi = \frac{4}{8} = \frac{1}{2}$$

$$\cos \frac{1}{7}\pi\ \cos \frac{3}{7}\pi + \cos \frac{1}{7}\pi \cos \frac{3}{7}\pi\ \cos \frac{5}{7}\pi = -\frac{4}{8} = -\frac{1}{2}$$

and $\cos \frac{1}{7}\pi\ \cos \frac{5}{7}\pi \cos \frac{3}{7}\pi = -\frac{1}{8}$

$$\therefore \quad \sec \frac{1}{7}\pi + \sec \frac{3}{7}\pi + \sec \frac{5}{7}\pi = \frac{1}{\cos \frac{1}{7}\pi} + \frac{1}{\cos \frac{3}{7}\pi} + \frac{1}{\cos \frac{5}{7}\pi}$$

$$= \frac{\cos\frac{3}{7}\pi \cos\frac{5}{7}\pi + \cos\frac{1}{7}\pi \cos\frac{5}{7}\pi + \cos\frac{1}{7}\pi \cos\frac{3}{7}\pi}{\cos\frac{1}{7}\pi \cos\frac{3}{7}\pi \cos\frac{5}{7}\pi}$$

$$= \frac{-1/2}{-1/8} = \frac{8}{2} = 4$$ **Ans.**

(e) $\cos\left(\frac{1}{7}\pi\right).\cos\left(\frac{3}{7}\pi\right).\cos\left(\frac{5}{7}\pi\right)$ = product of the roots of the equation (iv).

$$= \frac{1}{8}$$ **Ans.**

Example 17:

Prove that $\sin\frac{\pi}{14}$ *is a root of the equation*

$64z^6 - 80z^4 + 24z^2 - 1 = 0.$

Solution :

We have proved in that the equation whose roots are $\cos\frac{1}{7}\pi$, $\cos\frac{3}{7}\pi$ and $\cos\frac{5}{7}\pi$ is

$$8x^3 - 4x^2 - 4x + 1 = 0 \qquad ...(i)$$

Now as $\cos\frac{1}{7}\pi = 1 - 2\sin^2(\pi/14)$

$\therefore$ Put $x = 1 - 2z^2$ in equation (i) to obtain the required equation.

Hence the required equation is

$$8(1 - 2z^2)^3 - 4(1 - 2z^2)^2 - 4(1 - 2z^2) + 1 = 0$$

$$\Rightarrow 8(1 - 6z^2 + 12z^4 - 8z^6) - 4(1 - 4z^2 + 4z^4) - 4(1 - 2z^2) + 1 = 0$$

$$\Rightarrow 8 - 8z^2 + 96z^4 - 64z^6 - 4 + 16z^2 - 16z^4 - 4 + 8z^2 + 1 = 0$$

$$\Rightarrow 64z^6 - 80z^4 + 24z^2 - 1 = 0.$$ **Hence proved.**

Example 18:

Prove that $\cos\frac{2}{7}\pi$, $\cos\frac{4}{7}\pi$, $\cos\frac{8}{7}\pi$ *are the roots of the equation* $8x^3 + 4x^2 - 4x - 1 = 0.$

Solution :

Let θ denote one of the seven angles

$$\frac{2}{7}\pi, \frac{4}{7}\pi, \frac{6}{7}\pi, \frac{8}{7}\pi, \frac{10}{7}\pi, \frac{12}{7}\pi, \frac{14}{7}\pi \qquad \textbf{(Note)} \qquad ..(i)$$

i.e., $\theta = \frac{2}{7}n\pi$, where n = 1, 2, 3,..., 7

$\Rightarrow$ $7\theta = 2n\pi$ $\Rightarrow$ $4\theta = 2n\pi - 3\theta$

$\Rightarrow$ $\cos 4\theta = \cos(2n\pi - 3\theta) = \cos 3\theta$ **(Note)**

$\Rightarrow$ $8\cos^4\theta - 8\cos^2\theta + 1 = 4\cos^3\theta - 3\cos\theta$

$\Rightarrow$ $8x^4 - 4x^3 - 8x^2 + 3x + 1 = 0$, where $x = \cos\theta$

$\Rightarrow$ $(x - 1)(8x^3 + 4x^2 - 4x - 1) = 0$ ...(ii)

The roots of this equation are x = cow θ, where θ can take any one of the values given in (i).

Also $\cos\frac{14}{7}\pi = \cos 2\pi = 1$ and (x – 1) is the factor corresponding to the value $\frac{14}{7}\pi$ of θ. Hence the roots of the equation $8x^3 + 4x^3 - 4x - 1 = 0$ are the remaining six values of θ in (i). But we know that

$$\cos\frac{12}{7}\pi = \cos\left(2\pi - \frac{2}{7}\pi\right) = \cos\frac{2}{7}\pi; \cos\frac{10}{7}\pi = \cos\left(2\pi - \frac{4}{7}\pi\right) = \cos\frac{4}{7}\pi$$

and $\cos\frac{6}{7}\pi = \cos\left(2\pi - \frac{8}{7}\pi\right) = \cos\frac{8}{7}\pi$. ...(a)

Hence the roots of $8x^3 + 4x^2 - 4x - 1 = 0$ are $\cos\frac{2}{7}\pi$, $\cos\frac{4}{7}\pi$ and $\cos\frac{8}{7}\pi$. **Hence proved.**

Example 19:

In the above example find the equations whose rots are $sec^2\frac{2}{7}\pi, sec^2\frac{4}{7}\pi, sec^2\frac{6}{7}\pi$.

Solution :

From above we find that the equation whose roots are

$\cos\frac{2}{7}\pi, \cos\frac{4}{7}\pi, \cos\frac{6}{7}\pi$ or $\cos\frac{8}{7}\pi$ see. Result (a) of above] is

$$8x^3 + 4x^2 - 4x - 1 = 0 \qquad ...(i)$$

Let θ denote one of the angles $\frac{2}{7}\pi, \frac{4}{7}\pi, \frac{6}{7}\pi$, then $x = \cos\theta$ is solution of (i).

Now suppose $y = \sec^2\theta$

Then $y = 1/(\cos^2\theta) = 1/x^2 \Rightarrow x^2 = 1/y \Rightarrow x = 1/\sqrt{y}$

Substituting this value of x in (i) we get

$$8\left(1/\sqrt{y}\right)^3 + 4(1/\sqrt{y})^2 - 4(1/\sqrt{y}) - 1 = 0$$

$$\left(8/y\sqrt{y}\right) + (4/y) - \left(4/\sqrt{y}\right) - 1 = 0$$

$\Rightarrow \quad 8 + 4/\sqrt{y} - 4y - y\sqrt{y} = 0 \quad 8 - 4y = (y-4)\sqrt{y}$

$\Rightarrow \quad (8 - 4y)^2 = (y-4)^2 y \Rightarrow 64 - 64y + 16)^2 = y^3 - 8y^2 + 16y$

$\Rightarrow \quad y^3 - 24y^2 + 80y - 64 = 0$, which is the required equation whose roots are $\sec^2\frac{2}{7}\pi$, $\sec^2\frac{4}{7}\pi$, $\sec^2\frac{6}{7}\pi$.

Example 20:

Prove that $\sin\frac{2\pi}{7}$, $\sin\frac{4\pi}{7}$ *and* $\sin\frac{8\pi}{7}$ *are the roots of the equation* $x^3 - \frac{\sqrt{7}}{2}x^2 + \frac{\sqrt{7}}{8} = 0$.

Solution :

Let θ denote any one of the angles $\frac{2}{7}\pi, \frac{4}{7}\pi$ and $\frac{8}{7}\pi$.

Then $7\theta = 2n\pi$ an even multiple of π

$\Rightarrow \quad 4\theta = 2n\pi - 3\theta \Rightarrow \sin 4\theta = \sin(2n\pi - 3\theta) = -\sin 3\theta$

$\Rightarrow \quad 2\sin 2\theta\cos 2\theta = -(3\sin\theta - 4\sin^3\theta)$

$\Rightarrow \quad 4\sin\theta\cos\theta\cos 2\theta + 3\sin\theta - 4\sin^3\theta = 0$

$\Rightarrow \quad 4\cos\theta\cos 2\theta - 4\sin^2\theta + 3 = 0$, since $\sin\theta \neq 0$

$\Rightarrow \quad 4\cos\theta(1 - 2\sin^2\theta) - 4\sin^2\theta + 3 = 0$

$\Rightarrow \quad 4\sqrt{(1-\sin^2\theta)}\,(1 - 2\sin^2\theta) - 4\sin^2\theta + 3 = 0$

$\Rightarrow \quad 4\sqrt{(1-x^2)}\,(1 - 2x^2) = 4x^2 - 3$, where $x = \sin\theta$

Squaring both sides we have $16(1-x^2)(1-2x^2)^2 = (4x^2-3)^2$

$\Rightarrow \quad 64x^6 - 112x^4 + 56x^2 - 7 = 0 \qquad$ ...(i)

which being a cubic equation in x^2 has its roots as $\sin^2 2\pi/7$, $\sin^2 4\pi/7$ and $\sin^2 8\pi/7$

∴ From (i) we have

$$\sin^2 \frac{2\pi}{7} + \sin^2 \frac{4\pi}{7} + \sin^2 \frac{8\pi}{7} = \text{sum of the roots} = \frac{112}{64} = \frac{7}{4}$$

...(ii)

Also $\sin \frac{2}{7}\pi \sin \frac{4}{7}\pi + \sin \frac{4}{7}\pi \sin \frac{8}{7}\pi + \sin \frac{8}{7}\pi \sin \frac{2}{7}\pi$

$$= \frac{1}{2}\left[2 \sin \frac{2}{7}\pi \sin \frac{4}{7}\pi + 2 \sin \frac{8}{7}\pi \sin \frac{4}{7}\pi + 2 \sin \frac{8}{7}\pi \sin \frac{2}{7}\pi\right]$$

$$= \frac{1}{2}\left[\left(\cos \frac{2}{7}\pi - \cos \frac{6}{7}\pi\right) + \left(\cos \frac{4}{7}\pi - \cos \frac{12}{7}\pi\right) + \left(\cos \frac{6}{7}\pi - \sin \frac{10}{7}\pi\right)\right]$$

$$= \frac{1}{2}\left[\cos \frac{2}{7}\pi + \cos \frac{4}{7}\pi - \cos \frac{12}{7}\pi - \cos \frac{10}{7}\pi\right]$$

$$= \frac{1}{2}\left[\cos \frac{2}{7}\pi + \cos \frac{4}{7}\pi - \cos\left(2\pi - \frac{2}{7}\pi\right) - \cos\left(2\pi - \frac{4}{7}\pi\right)\right]$$

$$= \frac{1}{2}\left[\cos \frac{2}{7}\pi + \cos \frac{4}{7}\pi - \cos \frac{2}{7}\pi - \cos \frac{4}{7}\pi\right] = 0$$

$$\Rightarrow \sin \frac{2}{7}\pi \sin \frac{4}{7}\pi + \sin \frac{4}{7}\pi \sin \frac{8}{7}\pi + \sin \frac{8}{7}\pi \sin \frac{2}{7}\pi = 0 \qquad ...(iii)$$

Again from equation (i) we have

$$\sin^2 \frac{2}{7}\pi \sin^2 \frac{4}{7}\pi \sin^2 \frac{8}{7}\pi = \text{Product of the roots} = -(-7)/64 = 7/64$$

$$\Rightarrow \sin \frac{2\pi}{7} \sin \frac{4\pi}{7} \sin \frac{8\pi}{7} = -\sqrt{\frac{7}{64}} = -\frac{\sqrt{7}}{8} \qquad ...(iv)$$

[– ve value is taken as sin 8π/7 = – ve] **(Note)**

Now $\left(\sin \frac{2}{7}\pi + \sin \frac{4}{7}\pi + \sin \frac{8}{7}\pi\right)^2$

$$= \left(\sin^2 \frac{2}{7}\pi + \sin^2 \frac{4}{7}\pi + \sin^2 \frac{8}{7}\pi\right) + 2\left(\sin \frac{2}{7}\pi + \sin \frac{4}{7}\pi + ...\right)$$

= (7/4) + 2 (0), from (iii) and (ii)

$$\sin \frac{2}{7}\pi + \sin \frac{4}{7}\pi + \sin \frac{8}{7}\pi = \sqrt{\frac{7}{4}} = \frac{\sqrt{7}}{2} \qquad ...(v)$$

Hence the equation whose roots are

$\sin 2\pi/7$, $\sin 4\pi/7$ and $\sin 8\pi/7$ is

$x^3 - x^2$ (Sum of roots taken once

+ x (Sum of roots taken two at a time) – (Product of the roots) = 0

$$\Rightarrow \quad x^3 - x^2 \left(\sin\frac{2}{7}\pi + \sin\frac{4}{7}\pi + \sin\frac{8}{7}\pi\right)$$

$$+ x\left(\sin\frac{2}{7}\pi \sin\frac{4}{7}\pi + \sin\frac{4}{7}\pi \sin\frac{8}{7}\pi + \sin\frac{8}{7}\pi \sin\frac{2}{7}\pi\right)$$

$$-\left(\sin\frac{2\pi}{7} \sin\frac{4\pi}{7} \sin\frac{8\pi}{7}\right) = 0$$

$$\Rightarrow \quad x^3 - x^2\left(\sqrt{7/2}\right) + x(0) - \left(-\sqrt{7/8}\right) = 0$$

$$\Rightarrow \quad x^3 - \frac{1}{2}\sqrt{7x^2} + \frac{1}{8}\sqrt{7} = 0.$$

Hence proved.

Example 21:

Form the equation whose roots are

$$\tan^2\frac{\pi}{11}, \tan^2\frac{2\pi}{11}, \tan^2\frac{3\pi}{11}, \tan^2\frac{4\pi}{11} \text{ and } \tan^2\frac{5\pi}{11}.$$

Hence prove that

(a) $\sec^2\frac{\pi}{11} + \sec^2\frac{2\pi}{11} + \sec^2\frac{3\pi}{11} + \sec^2\frac{4\pi}{11} + \sec^2\frac{5\pi}{11} = 60$

and (b) $\cot^2\frac{\pi}{11} + \cot^2\frac{2\pi}{11} + \cot^2\frac{3\pi}{11} + \cot^2\frac{4\pi}{11} + \cot^2\frac{5\pi}{11} = 15$.

Solution :

Let θ be any one of the angles

$$\frac{\pi}{11}, \frac{2\pi}{11}, \frac{3\pi}{11}, \frac{4\pi}{11}, \frac{5\pi}{11}, \frac{6\pi}{11}, \frac{7\pi}{11}, \frac{8\pi}{11}, \frac{9\pi}{11}, \frac{10\pi}{11} \text{ and } \frac{11\pi}{11}$$

Then $\tan 11\theta = 0$, ∴ $\tan n\pi = 0$, for all integral values of n

$$\Rightarrow \frac{11c_1 \tan\theta - 11c_3 \tan^3\theta + 11c_5 \tan^5\theta - 11c_7 \tan^7\theta + 11c_9 \tan^9\theta - 11c_{11} \tan^{11}\theta}{1 - 11c_2 \tan^2\theta + 11c_4 \tan^4\theta - 11c_6 \tan^6\theta + 11c_8 \tan^8\theta - 11c_{10} \tan^{10}\theta} = 0$$

$$\Rightarrow 11c_1 \tan\theta - 11c_3 \tan^3\theta + 11c_5 \tan^5\theta - 11c_7 \tan^7\theta + 11c_9 \tan^9\theta - 11c_{11} \tan^{11}\theta = 0$$

$\Rightarrow$ $11 \tan\theta - 165 \tan^3\theta + 462 \tan^5\theta - 330 \tan^7\theta + 55 \tan^9\theta - \tan^{11}\theta = 0,$

putting the values of $11c_1$, $11c_3$ etc.

$\Rightarrow$ $11x - 165x^3 + 462x^5 - 330x^7 + 55x^9 - x^{11} = 0$, where $x = \tan\theta$

$\Rightarrow$ $x(x^{10} - 55x^8 + 330x^6 - 462x^4 + 165x^2 - 11) = 0$

But $x = 0$ corresponds to the value of $\theta = \dfrac{11\pi}{11} = \pi$

Hence the equation

$$x^{10} - 55x^8 + 330x^6 - 462x^4 + 165x^2 - 11 = 0 \qquad \text{...(i)}$$

has its roots as tan $\tan\dfrac{\pi}{11}, \tan\dfrac{2\pi}{11}, \tan\dfrac{3\pi}{11}, \ldots, \tan\dfrac{10\pi}{11}$

But this is a fifth degree equation in x^2.

$\therefore$ putting $x^2 = y$, this reduces to

$$y^5 - 55y^4 - 330y^3 - 462y^2 + 165y - 11 = 0 \qquad \text{...(ii)}$$

whose roots are

$$\tan^2\frac{\pi}{11}, \tan^2\frac{2\pi}{11}, \tan^2\frac{3\pi}{11}, \tan^2\frac{4\pi}{11} \text{ and } \tan^2\frac{5\pi}{11}$$

(a) From equation (ii) above we find that

$$\tan^2\frac{\pi}{11} + \tan^2\frac{2\pi}{11} + \tan^2\frac{3\pi}{11} + \tan^2\frac{4\pi}{11} + \tan^2\frac{5\pi}{11}$$

$$= \text{sum of the roots taken once} = \frac{55}{1} = 55$$

Adding 5 to both sides we have

$$\left(1+\tan^2\frac{\pi}{11}\right) + \left(1+\tan^2\frac{2\pi}{11}\right) + \left(1+\tan^2\frac{3\pi}{11}\right) + \left(1+\tan^2\frac{4\pi}{11}\right) + \left(1+\tan^2\frac{5\pi}{11}\right) = 55 + 5 \quad \textbf{(Note)}$$

$$\Rightarrow \sec^2\frac{\pi}{11} + \sec^2\frac{2\pi}{11} + \sec^2\frac{3\pi}{11} + \sec^2\frac{4\pi}{11} + \sec^2\frac{5\pi}{11} = 60.$$

Hence proved.

(b) We have proved above that the equation (ii) viz.,

$y^5 - 55y^4 + 330y^3 - 462y^2 + 165y - 11 = 0$ has its roots

as $\tan^2\dfrac{\pi}{11}, \tan^2\dfrac{2\pi}{11}, \tan^2\dfrac{3\pi}{11}, \tan^2\dfrac{4\pi}{11}$ and $\tan^2\dfrac{5\pi}{11}$

Also ∴ $\cot^2(\pi/11) = 1/\tan^2(\pi/11)$

∴ putting $x = 1/y \Rightarrow y = 1/x$ we have the equation whose roots are

$$\cot^2\frac{\pi}{11}, \cot^2\frac{2\pi}{11}, \ldots, \cot^2\frac{5\pi}{11} \text{ as}$$

$$\frac{1}{x^5} - \frac{55}{x^4} + \frac{330}{x^3} - \frac{462}{x^2} + \frac{165}{x} - 11 = 0$$

$$\Rightarrow 11x^5 - 165x^4 + 462x^3 - 330x^2 + 55x - 1 = 0,$$

whence $\cot^2\frac{\pi}{11} + \cot^2\frac{2\pi}{11} + \cot^2\frac{3\pi}{11} + \cot^2\frac{4\pi}{11} + \cot^2\frac{5\pi}{11}$

= sum of the roots taken once $= -(-165)/11 = 15$

Hence proved.

Example 22:

Express the following in the form a + ib :

(a) $\dfrac{(\cos 3\theta + i\sin 3\theta)^5}{(\cos\theta + i\sin\theta)^6}$ (b) $\dfrac{(\cos 3\alpha + i\sin 3\alpha)^{10}}{(\cos 2\alpha + i\sin 2\alpha)^4}$.

Solution :

(a) By De Moivre's Theorem, we have

$$\frac{(\cos 3\theta + i\sin 3\theta)^5}{(\cos\theta + i\sin\theta)^6} = \frac{\cos(5\times 3\theta) + i\sin(5\times 3\theta)}{\cos 6\theta + i\sin 6\theta}$$

$$= \frac{\cos 15\theta + i\sin 15\theta}{\cos 6\theta + i\sin 6\theta}$$

$= (\cos 15\theta + i\sin 15\theta)(\cos 6\theta - i\sin 6\theta)$ [Remark 1]

$= \cos(15\theta - 6\theta) + i\sin(15\theta - 6\theta)$ [Remark 3]

$= \cos 9\theta + i\sin 9\theta.$

(b) $$\frac{(\cos\alpha + i\sin\alpha)^{10}}{(\cos 2\alpha + i\sin 2\alpha)^4} = \frac{\cos 10\alpha + i\sin 10\alpha}{\cos 8\alpha + i\sin 8\alpha}$$

$= \cos(10\alpha - 8\alpha) + i\sin(10\alpha - 8\alpha)$

$= \cos 2\alpha + i\sin 2\alpha.$

Example 23:

Find the modulus and the argument of the complex number

$$z = \frac{(\sin\alpha + i\ coa\ \alpha)^4}{(\cos\alpha - i\sin\alpha)^3}.$$

Solution :

$(\sin\alpha + i\cos\alpha)^4 = \{\cos(\pi/2 - \alpha) + i\sin(\pi/2 - \alpha)\}^4$

$= \cos(2\pi - 4\alpha) + i\sin(2\pi - 4\alpha)$ (De Moivre's Theorem)

Also $(\cos\alpha - i\sin\alpha)^3 = \cos 3\alpha - i\sin 3\alpha$.

$\therefore\ z = \{\cos(2\pi - 4\alpha) + i\sin(2\pi - 4\alpha)\}\{\cos 3\alpha + i\sin 3\alpha\}$

$= \cos\{(2\pi - 4\alpha) + 3\alpha\} + i\sin\{(2\pi - 4\alpha) + 3\alpha\}$

$= \cos(2\pi - \alpha) + i\sin(2\pi - \alpha)$

$= \cos\alpha - i\sin\alpha = \cos(-\alpha) + i\sin(-\alpha)$.

Hence the modulus of z is $\sqrt{\cos^2\alpha + \sin^2\alpha = 1}$ and the argument of z is $-\alpha$.

Example 24:

Find the modulus and the argument of the complex number $\frac{(1 + \cos\theta + i\sin\theta)^5}{(\cos\theta + i\sin\theta)^3}$.

Solution :

$$\text{Let } z = \frac{(1 + \cos\theta + i\sin\theta)^5}{(\cos\theta + i\sin\theta)^3}$$

$$= \frac{\{1 + 2\cos^2(\theta/2) - 1 + 2i\sin\theta/2\cos\theta/2\}^5}{\cos 3\theta + i\sin 3\theta}$$

$$= \frac{32\cos^5(\theta/2)\{\cos\theta/2 + i\sin\theta/2\}^5}{\cos 3\theta + i\sin 3\theta}$$

$= 32\cos^5(\theta/2)\{\cos(5\theta/2) + i\sin(5\theta/2)\}(\cos 3\theta - i\sin 3\theta)$

$= 32\cos^5(\theta/2)[\cos\{(5\theta/2) - 3\theta\} + i\sin\{(5\theta/2) - 3\theta\}]$

$= 32\cos^5(\theta/2)[\cos(-\theta/2) + i\sin(-\theta/2)]$.

Hence, the modulus and argument of z is $32|\cos^5(\theta/2)|$ and $(-\theta/2)$ respectively.

Example 25:

if $z = \cos\theta + i\sin\theta$, *show that*

$$z^n + \frac{1}{z^n} = 2\ \cos n\theta,\ z^n - \frac{1}{z^n} = 2i\ \sin n\theta.$$

Solution :

We have $\frac{1}{z} = \frac{1}{\cos\theta + i\sin\theta} = \cos\theta - i\sin\theta.$

$\therefore\quad z^n = (\cos\theta + i\sin\theta)^n = \cos n\theta + i\sin n\theta,$

and $\frac{1}{z^n}\ (\cos\theta - i\sin\theta)^n = \cos n\theta - i\sin n\theta.$

Hence $z^n + \frac{1}{z^n} = 2\cos n\theta$ and $z^n - \frac{1}{z^n} = 2i\ \sin n\theta.$

Example 26:

If $z = \cos\theta + i\sin\theta$, show that

$\frac{z^{2n}-1}{z^{2n}+1} = i\ \tan n\theta$, *n being an integer.*

Solution :

We have

$$\frac{z^{2n}-1}{z^{2n}+1} = \frac{(\cos\theta + i\sin\theta)^{2n} - 1}{(\cos\theta + i\sin\theta)^{2n} - 1}$$

$$= \frac{\cos 2n\theta + i\sin 2n\theta - 1}{\cos 2n\theta + i\sin 2n\theta + 1} \text{ (De Moivre's Theorem)}$$

$$= \frac{(1 - 2\sin^2 n\theta) + 2i\sin n\theta\cos n\theta - 1}{(2\cos^2 n\theta - 1) + 2i\sin n\theta\cos n\theta + 1}$$

$$= \frac{i\sin n\theta\cos n\theta + i^2\sin^2 n\theta}{\cos^2 n\theta + i\sin n\theta\cos n\theta} \qquad (\therefore\ i^2 = -1)$$

$$= \frac{i\sin n\theta(\cos n\theta + i\sin n\theta)}{\cos n\theta(\cos n\theta + i\sin n\theta)} = i\tan n\theta.$$

Example 27:

If n is an integer, show that

$$(1 + \cos\theta + i\sin\theta)^n + (1 + \cos\theta - i\sin\theta)^n = 2^{n+1}\cos^n\frac{\theta}{2}\cos\frac{n\theta}{2}.$$

Solution :

$(1 + \cos\theta + i\sin\theta)^n + (1 + \cos\theta - i\sin\theta)^n$

$= [1 + 2\cos^2\theta/2 - 1 + 2i\sin\theta/2\cos\theta/2]^n$
$\quad + [1 + 2\cos^2\theta/2 - 1 - 2i\sin\theta/2\cos\theta/2]^n$

$= 2^n\cos^n\theta/2\,[\cos\theta/2 + i\sin\theta/2]^n$
$\quad + 2^n\cos^n\theta/2\,[\cos\theta/2 - i\sin\theta/2]^n$

$$= 2^n\cos^n\frac{\theta}{2}\left[\left(\cos\frac{n\theta}{2} + i\sin\frac{n\theta}{2}\right) + \left(\cos\frac{n\theta}{2} - i\sin\frac{n\theta}{2}\right)\right]$$

[By De Moivre's Theorem]

$$= 2^n\cos^n\frac{\theta}{2}.2\cos\frac{n\theta}{2} = 2^{n+1}\cos^n\frac{\theta}{2}\cos\frac{n\theta}{2}.$$

Example 28:

Show that

(i) $(\sin\theta + i\cos\theta)^n = \cos n\left(\frac{\pi}{2} - \theta\right) + i\sin n\left(\frac{\pi}{2} - \theta\right)$, *n being an integer.*

(ii) $\dfrac{1 + \cos\theta + i\sin\theta}{1 + \cos\theta - i\sin\theta} = \cos\theta + i\sin\theta.$

Solution :

We have

(i) $(\sin\theta + i\cos\theta)^n = \left[\cos\left(\frac{\pi}{2} - \theta\right) + i\sin\left(\frac{\pi}{2} - \theta\right)\right]^n$

$$= \cos n\left(\frac{\pi}{2} - \theta\right) + i\sin n\left(\frac{\pi}{2} - \theta\right).$$

(ii) $\dfrac{1 + \cos\theta + i\sin\theta}{1 + \cos\theta - i\sin\theta} = \dfrac{1 + 2\cos^2\theta/2 - 1 + 2i\sin\theta/2\cos\theta/2}{1 + 2\cos^2\theta/2 - 1 - 2i\sin\theta/2\cos\theta/2}$

$$= \frac{\cos\theta/2\,[\cos\theta/2 + i\sin\theta/2]}{\cos\theta/2\,[\cos\theta/2 - i\sin\theta/2]}$$

$$= \frac{(\cos\theta/2 + \sin\theta/2)(\cos\theta/2 + i\sin\theta/2)}{(\cos\theta/2 - i\sin\theta/2)(\cos\theta/2 + i\sin\theta/2)}$$

$= (\cos\theta/2 + i\sin\theta/2)^2 = \cos\theta + i\sin\theta$, (by De Moivre's Theorem)

Example 29:

We have

$$\left[\frac{1+\sin\phi+i\cos\phi}{1+\sin\phi-i\cos\phi}\right]=\cos\left(\frac{n\pi}{2}-n\phi\right)+i\sin\left(\frac{n\pi}{2}-n\phi\right).$$

Solution :

We have

$$\left[\frac{1+\sin\phi+i\cos\phi}{1+\sin\phi-i\cos\phi}\right]=\left[\frac{1+\cos\left(\frac{\pi}{2}-\phi\right)+i\sin\left(\frac{\pi}{2}-\phi\right)}{1+\cos\left(\frac{\pi}{2}-\phi\right)-\sin\left(\frac{\pi}{2}-\phi\right)}\right]$$

$$=\left[\frac{1+2\cos^2\left(\frac{\pi}{4}-\frac{\phi}{2}\right)-1+2i\sin\left(\frac{\pi}{4}-\frac{\phi}{2}\right)\cos\left(\frac{\pi}{4}-\frac{\phi}{2}\right)}{1+2\cos^2\left(\frac{\pi}{4}-\frac{\phi}{2}\right)-1-2i\sin\left(\frac{\pi}{4}-\frac{\phi}{2}\right)\cos\left(\frac{\pi}{4}-\frac{\phi}{2}\right)}\right]^n$$

$$=\frac{2^n\cos^n\left(\frac{\pi}{4}-\frac{\phi}{2}\right)\left\{\cos\left(\frac{\pi}{4}-\frac{\phi}{2}\right)+i\sin\left(\frac{\pi}{4}-\frac{\phi}{2}\right)\right\}^n}{2^n\cos^n\left(\frac{\pi}{4}-\frac{\phi}{2}\right)\left\{\cos\left(\frac{\pi}{4}-\frac{\phi}{2}\right)-i\sin\left(\frac{\pi}{4}-\frac{\phi}{2}\right)\right\}^n}$$

$$=\frac{\cos n\left(\frac{\pi}{4}-\frac{\phi}{2}\right)+i\sin n\left(\frac{\pi}{4}-\frac{\phi}{2}\right)}{\cos n\left(\frac{\pi}{4}-\frac{\phi}{2}\right)-i\sin n\left(\frac{\pi}{4}-\frac{\phi}{2}\right)}\quad\text{(De Moivre's Theorem)}$$

$$=\left[\cos n\left(\frac{\pi}{4}-\frac{\phi}{2}\right)+i\sin n\left(\frac{\pi}{4}-\frac{\phi}{2}\right)\right]\times\left[\cos n\left(\frac{\pi}{4}-\frac{\phi}{2}\right)+i\sin n\left(\frac{\pi}{4}-\frac{\phi}{2}\right)\right]$$

$$=\left\{\cos n\left(\frac{\pi}{4}-\frac{\phi}{2}\right)+i\sin n\left(\frac{\pi}{4}-\frac{\phi}{2}\right)\right\}^2$$

$$=\cos\left(\frac{n\pi}{2}-n\phi\right)+i\sin\left(\frac{n\pi}{2}-n\phi\right).\quad\text{[De Moivre's Theorem]}$$

Example 30:

Express the following in the form $x + iy$

$$\left[\frac{1 + \sin\theta + i\cos\theta}{1 + \sin\theta - i\cos\theta}\right]^n$$

Solution :

It is same as Example where

$$x + iy = \cos\left(\frac{n\pi}{2} - n\theta\right) + i\sin\left(\frac{n\pi}{2} - n\theta\right).$$

Example 31:

Find the modulus and argument of

$$\left(\frac{1 + \cos\theta + i\sin\theta}{1 + \cos\theta - i\sin\theta}\right)^n,$$

n is a positive integer and $\theta \neq (2k + 1)\pi$.

Solution :

$$\text{Let } z = \frac{1 + \cos\theta + i\sin\theta}{1 + \cos\theta - i\sin\theta}$$

$$= \frac{1 + 2\cos^2(\theta/2) - 1 + 2i\sin\theta/2\cos\theta/2}{1 + 2\cos^2(\theta/2) - 1 - 2i\sin\theta/2\cos\theta/2}$$

$$= \frac{\cos\theta/2 + i\sin\theta/2}{\cos\theta/2 - i\sin\theta/2}, \text{ as } \cos\theta/2 \neq 0$$

$$= (\cos\theta/2 + i\sin\theta/2)(\cos\theta/2 + i\sin\theta/2)$$

$$= (\cos\theta/2 + i\sin\theta/2)^2$$

$$\therefore z^n = (\cos\theta/2 + i\sin\theta/2)^{2n} = \cos n\theta + i\sin n\theta$$

(De Moivre's Theorem)

Hence the modulus of z^n is $\sqrt{\cos^2 n\theta + \sin^2 n\theta} = 1$ and the argument of z^n is $n\theta$.

Example 32:

Show that

$$\{(\cos\alpha + i\sin\alpha) + (\cos\beta + i\sin\beta)\}^n + \{(\cos\alpha - i\sin\alpha) + (\cos\beta - i\sin\beta)\}^n$$

$$= 2^{n+1}\cos^n\left(\frac{\alpha - \beta}{2}\right)\cos\left\{\frac{n}{2}(\alpha + \beta)\right\},$$ *n being an integer.*

Solution :

L.H.S. = $\{(\cos\alpha + \cos\beta) + i(\sin\alpha + \sin\beta)\}^n$

$+ \{(\cos\alpha + \cos\beta) - i(\sin\alpha + \sin\beta)\}^n$

$$= \left\{2\cos\frac{\alpha+\beta}{2}\cos\frac{\alpha-\beta}{2} + 2i\sin\frac{\alpha+\beta}{2}\cos\frac{\alpha-\beta}{2}\right\}^n$$

$$+ \left\{2\cos\frac{\alpha+\beta}{2}\cos\frac{\alpha-\beta}{2} - 2i\sin\frac{\alpha+\beta}{2}\cos\frac{\alpha-\beta}{2}\right\}^n$$

$$= 2^n\cos^n\left(\frac{\alpha-\beta}{2}\right)\left\{\cos\frac{\alpha+\beta}{2} + i\sin\frac{\alpha+\beta}{2}\right\}^n$$

$$+ 2^n\cos^n\left(\frac{\alpha-\beta}{2}\right)\left\{\cos\frac{\alpha+\beta}{2} - i\sin\frac{\alpha+\beta}{2}\right\}^n$$

$$= 2^n\cos^n\left(\frac{\alpha-\beta}{2}\right)\left[\left\{\cos\frac{n(\alpha+\beta)}{2} + i\sin\frac{n(\alpha+\beta)}{2}\right\}\right.$$

$$\left. + \left\{\cos\frac{n(\alpha+\beta)}{2} - i\sin\frac{n(\alpha+\beta)}{2}\right\}\right] \quad \text{(by De Moivre's Theorem)}$$

$$= 2^n\cos^n\left(\frac{\alpha-\beta}{2}\right)\left[2\cos\frac{n(\alpha+\beta)}{2}\right]$$

$$= 2^{n+1}\cos^n\left(\frac{\alpha-\beta}{2}\right)\cos\left\{\frac{n}{2}(\alpha+\beta)\right\}$$

Example 33:

If $m = \cos\alpha + i\sin\alpha$ and $n = \cos\beta + i\sin\beta$, prove that

$$\frac{m-n}{m+n} = i\tan\left(\frac{\alpha-\beta}{2}\right).$$

Solution :

We have

$$m - n = (\cos\alpha - \cos\beta) + i(\sin\alpha - \sin\beta)$$

$$= -2\sin\frac{\alpha+\beta}{2}\sin\frac{\alpha-\beta}{2} + 2i\cos\frac{\alpha+\beta}{2}\sin\frac{\alpha-\beta}{2}$$

$$\therefore\ m - n = 2i\sin\frac{\alpha-\beta}{2}\left[\cos\frac{\alpha+\beta}{2} + i\sin\frac{\alpha+\beta}{2}\right]$$

Now $m + n = (\cos\alpha + \cos\beta) + i(\sin\alpha + \sin\beta)$ ...(1)

$$= 2\cos\frac{\alpha+\beta}{2}\cos\frac{\alpha-\beta}{2} + 2i\sin\frac{\alpha+\beta}{2}\cos\frac{\alpha-\beta}{2}$$

$$\therefore\quad m + n = 2\cos\frac{\alpha-\beta}{2}\left[\cos\frac{\alpha+\beta}{2} + i\sin\frac{\alpha+\beta}{2}\right]. \qquad ...(2)$$

From (1) and (2), we obtain

$$\frac{m+n}{m+n} = i\tan\left(\frac{\alpha-\beta}{2}\right).$$

Example 34:

If $x_r = \cos\frac{\pi}{2^r} + i\sin\frac{\pi}{2^r}$, *prove that*

$x_1x_2x_3 \ldots$ *to* $\infty = -1$.

Solution :

We have

$$x_1x_2x_2\ldots\infty = \left(\cos\frac{\pi}{2} + i\sin\frac{\pi}{2}\right)\left(\cos\frac{\pi}{2^2} + i\sin\frac{\pi}{2^2}\right)$$

$$\left(\cos\frac{\pi}{2^3} + i\sin\frac{\pi}{2^3}\right)\ldots\ldots$$

$$= \cos\left(\frac{\pi}{2} + \frac{\pi}{2^3} + \frac{\pi}{2^3} + \ldots\ldots\right) + i\sin\left(\frac{\pi}{2} + \frac{\pi}{2^3} + \frac{\pi}{2^3} + \ldots\ldots\right)$$

[Remark 4, p. 2] ...(1)

We note that $\frac{\pi}{2} + \frac{\pi}{2^2} + \frac{\pi}{2^3} + \ldots\ldots$ is a G.P., whose common ratio is r = 1/2 and the first term is α = π/2. Thus

$$\frac{\pi}{2} + \frac{\pi}{2^2} + \frac{\pi}{2^3} + \ldots\ldots = \frac{a}{1-r} = \frac{\pi/2}{1-\frac{1}{2}} = \pi\cdot$$

From (1), $x_1x_2x_3\ldots\infty = \cos\pi + i\sin\pi = -1$.

Example 35:

If $\sin\theta + \sin\phi + \sin\psi = 0 = \cos\theta + \cos\phi + \cos\psi$, show that

$\sin 3\theta + \sin 3\phi + \sin 3\psi = 3\sin(\theta + \phi + \psi)$.

and $\cos 3\theta + \cos 3\phi + \cos 3\psi = 3\cos(\theta + \phi + \psi)$.

Solution :

Let $x = \cos\theta + i\sin\theta$, $y = \cos\phi + i\sin\phi$, $z = \cos\psi + i\sin\psi$.

Now $x + y + z = (\cos\theta + \cos\phi + \cos\psi) + i(\sin\theta + \sin\phi + \sin\psi)$

$\Rightarrow\ x + y + z = 0$

$\Rightarrow x^3 + y^3 + z^3 = 3xyz.$

Thus $(\cos\theta + i\sin\theta)^3 + (\cos\phi + i\sin\phi)^3 + (\cos\psi + i\sin\psi)^3$

$= 3(\cos\theta + i\sin\theta)(\cos\phi + i\sin\phi)(\cos\psi + i\sin\psi)$

$\Rightarrow\ (\cos 3\theta + i\sin 3\theta) + (\cos 3\phi + i\sin 3\phi) + (\cos 3\psi + i\sin 3\psi)$

$= 3\{\cos(\theta + \phi + \psi) + i\sin(\theta + \phi + \psi)\}.$ [Remark 4]

Comparing real and imaginary parts, we get

$\cos 3\theta + \cos 3\phi + \cos 3\psi = 3\cos(\theta + \phi + \psi),$

and $\sin 3\theta + \sin 3\phi + \sin 3\psi = 3\sin(\theta + \phi + \psi).$

Example 36:

If $\cos\alpha + 2\cos\beta + 3\cos\gamma = \sin\alpha + 2\sin\beta + 3\sin\gamma = 0$, prove that

$\cos 3\alpha + 8\cos 3\beta + 27\cos 3\gamma = 18\cos(\alpha + \beta + \gamma),$

$\sin 3\alpha + 8\sin 3\beta + 27\sin 3\gamma = 18\sin(\alpha + \beta + \gamma).$

Solution :

Let $a = \cos\alpha + i\sin\alpha$, $b = 2\cos\beta + 2i\sin\beta$

and $c = 3\cos\gamma + 3i\sin\gamma$. Then

$a + b + c = (\cos\alpha + 2\cos\beta + 3\cos\gamma) + i(\sin\alpha + 2\sin\beta + 3\sin\gamma) = 0$

We know $a + b + c = 0 \Rightarrow a^3 + b^3 + c^3 = 3abc$

$\Rightarrow\ (\cos\alpha + i\sin\alpha)^3 + (2\cos\beta + 2i\sin\beta)^3 + (3\cos\gamma + 3i\sin\gamma)^3$

$= 3(\cos\alpha + i\sin\alpha)(2\cos\beta + 2i\sin\beta)(3\cos\gamma + 3i\cos\gamma)$

$\Rightarrow\ (\cos 3\alpha + i\sin 3\alpha) + 8(\cos 3\beta + i\sin 3\beta) + 27(\cos 3\gamma + i\sin 3\gamma)$

$= 18(\cos\alpha + i\sin\alpha)(\cos\beta + i\sin\beta)(\cos\gamma + i\sin\gamma)$

$\Rightarrow\ (\cos 3\alpha + 8\cos 3\beta + 27\cos 3\gamma) + i(\sin 3\alpha + 8\sin 3\beta + 27\sin 3\gamma)$

$= 18[\cos(\alpha + \beta + \gamma) + i\sin(\alpha + \beta + \gamma)].$

Comparing real and imaginary parts, we get

$\cos 3\alpha + 8 \cos 3\beta + 27 \cos 3\gamma = 18 \cos (\alpha + \beta + \gamma)$,

and $\sin 3\alpha + 8 \sin 3\beta + 27 \sin 3\gamma = 18 \sin (\alpha + \beta + \gamma)$.

Example 37:

If $a = \cos \alpha + i \sin \alpha$, $b = \cos \beta + i \sin \beta$,

$c = \cos \gamma + i \sin \gamma$, *and* $\frac{a}{b}+\frac{b}{c}+\frac{c}{a}=-1$, *prove that*

$\cos (\beta - \gamma) + \cos (\gamma - \alpha) + \cos (\alpha - \beta) + 1 = 0$.

Solution :

We have $1/a = \cos \alpha - i \sin \alpha$, $1/b = \cos \beta - i \sin \beta$.

Now $a/b = (\cos \alpha + i \sin \alpha)(\cos \beta - i \sin \beta)$

$\Rightarrow$ $a/b = \cos (\alpha - \beta) + i \sin (\alpha - \beta)$.

Similarly $b/c = \cos (\beta - \gamma) + i \sin (\beta - \gamma)$,

and $c/a = \cos (\gamma - \alpha) + i \sin (\gamma - \alpha)$.

Putting these values in $\frac{a}{b}+\frac{b}{c}+\frac{c}{a}=-1$, we get

$$[\cos (\alpha - \beta) + \cos (\beta - \gamma) + \cos (\gamma - \alpha)] + i [\sin (\alpha - \beta) + \sin (\beta - \gamma) + \sin (\gamma - \alpha)] = -1 = -1 + 0\,i.$$

Comparing real and imaginary parts, we get

$$\cos (\alpha - \beta) + \cos (\beta - \gamma) + \cos (\gamma - \alpha) = -1.$$

Hence $\cos (\alpha - \beta) + \cos (\beta - \gamma) + \cos (\gamma - \alpha) + 1 = 0$.

Example 38:

If $a \cos \alpha + i \sin \alpha$, $b = \cos \beta + i \sin \beta$,

$c = \cos \gamma + i \sin \alpha$, *prove that* $(b + c)(c + a)(a + b)/abc$

$$= 8 \cos \frac{1}{2} (\beta - \gamma) \cos \frac{1}{2} (\gamma - \alpha) \cos \frac{1}{2} (\alpha - \beta).$$

Solution :

$$\frac{(b+c)(c+a)(a+b)}{abc} = \left(\frac{b+c}{c}\right)\left(\frac{c+a}{a}\right)\left(\frac{a+b}{b}\right)$$

$$= \left(1+\frac{a}{b}\right)\left(1+\frac{b}{c}\right)\left(1+\frac{c}{a}\right). \qquad \text{...(1)}$$

By Example 16, $a/b = \cos (\alpha - \beta) + i \sin (\alpha - \beta)$

$$\therefore \left(1+\frac{a}{b}\right) = 1+2\cos^2\frac{1}{2}(\alpha-\beta) - 1 + 2i\sin\frac{1}{2}(\alpha-\beta)\cos\frac{1}{2}(\alpha-\beta)$$

$$= 2\cos\frac{1}{2}(\alpha-\beta)\left\{\cos\frac{1}{2}(\alpha-\beta) + i\sin\frac{1}{2}(\alpha-\beta)\right\} \text{ etc.}$$

...(2)

From (1) and (2), we obtain

$$\frac{(b+c)(c+a)(a+b)}{abc} = 8\cos\frac{1}{2}(\alpha-\beta)\cos\frac{1}{2}(\beta-\gamma)\cos\frac{1}{2}(\gamma-\alpha)$$

$$\times\left[\cos\left\{\frac{1}{2}(\alpha-\beta)+\frac{1}{2}(\beta-\gamma)+\frac{1}{2}(\gamma-\alpha)\right\}\right.$$

$$\left.+\sin\left\{\frac{1}{2}(\alpha-\beta)+\frac{1}{2}(\beta-\gamma)+\frac{1}{2}(\gamma-\alpha)\right\}\right]$$

$$= 8\cos\frac{1}{2}(\alpha-\beta)\cos\frac{1}{2}(\beta-\gamma)\cos\frac{1}{2}(\gamma-\alpha).$$

$(\because \cos 0 + i\sin 0 = 1)$

Example 39:

If $a = \cos 2\alpha + i\sin 2\alpha$, $b = \cos 2\beta + \sin 2\beta$, $c = \cos 2\gamma + i\sin 2\gamma$, $d = \cos 2\delta + i\sin 2\delta$, *show that*

$$\sqrt{abcd} + \frac{1}{\sqrt{abcd}} = 2\cos(\alpha+\beta+\gamma+\delta).$$

Solution :

We have

$$abcd = \cos(2\alpha+2\beta+2\gamma+2\delta) + i\sin(2\alpha+2\beta+2\gamma+2\delta)$$

$$\therefore \sqrt{abcd} = [\cos(2\alpha+2\beta+2\gamma+2\delta) + i\sin(2\alpha+2\beta+2\gamma+2\delta)]^{1/2}$$

$$\Rightarrow \sqrt{abcd} = \cos(\alpha+\beta+\gamma+\delta) + i\sin(\alpha+\beta+\gamma+\delta).$$

...(1)

[De Moivre's Theorem]

$$\therefore \frac{1}{\sqrt{abcd}} = \cos(\alpha+\beta+\gamma+\delta) - i\sin(\alpha+\beta+\gamma+\delta) \quad ....(2)$$

Adding (1) and (2), we obtain

$$\sqrt{abcd} + \frac{1}{\sqrt{abcd}} = 2\cos(\alpha+\beta+\gamma+\delta).$$

Example 41:

If $\cos\alpha + \cos\beta + \cos\gamma = \sin\alpha + \sin\beta + \sin\gamma = 0.$

then $\Sigma \cos 4\alpha = 2\Sigma \cos 2\alpha\,(\beta + \gamma)$, $\Sigma \sin 4\alpha = 2\Sigma \sin 2\,(\beta + \gamma)$.

Solution :

Let $a = \cos\alpha + i \sin\alpha$, $b = \cos\beta + i\sin\beta$

$c = \cos\gamma + i\sin\gamma$.

Then $a + b + c = 0 \Rightarrow a + b = -c \Rightarrow (a + b)^2 = c^2$

$\Rightarrow a^2 + b^2 - c^2 = -2ab \Rightarrow (a^2 + b^2 - c^2)^2 = 4a^2b^2$

$\Rightarrow a^4 + b^4 + c^4 = 2\,(a^2b^2 + b^2c^2 + c^2a^2)$. ...(1)

Now $a^2 = \cos 2\alpha + i\sin 2\alpha$ and $a^4 = \cos 4\alpha + i \sin 4\alpha$ etc.

We have $a^4 + b^4 + c^4 = (\cos 4\alpha + i\sin 4\alpha) + (\cos 4\beta + i \sin 4\beta) + (\cos 4\gamma + i\sin 4\gamma)$

$= \Sigma \cos 4\alpha + i\, S \sin 4\alpha.$ ...(2)

Now $2\,(b^2c^2 + c^2a^2 + a^2b^2) = 2\,(\cos 2\beta + i\sin 2\beta)(\cos 2\gamma + i \sin 2\gamma) + \ldots\ldots + \ldots\ldots$

$= 2\,[\cos 2\,(\beta + \gamma) + i\sin 2\,(\beta + \gamma)] + \ldots\ldots + \ldots\ldots$

$= 2\,\Sigma \cos 2\,(\beta + \gamma) + i\,2\,\Sigma\sin 2\,(\beta + \gamma).$ ...(3)

Using (2) and (3) in (1), we obtain

$\Sigma\cos 4\alpha + i\,\Sigma \sin 4\alpha = 2\,\Sigma\cos 2\,(\beta + \gamma) + i\,2\,\Sigma \sin 2\,(\beta + \gamma).$

Comparing real and imaginary parts, we get

$\Sigma \cos 4\alpha = 2\,\Sigma\cos 2\,(\beta + \gamma)$,

and $\Sigma\sin 4\alpha = 2\,\Sigma \sin 2\,(\beta + \gamma)$.

Example 42:

If $a = \cos\alpha + i\sin\alpha$, $b = \cos\beta + i\sin\beta$,

$c = \cos\gamma + i\sin\gamma$ *and* $d = \cos\delta + i\sin\delta$; *prove that*

(i) $(a + b)(c + d)$

$$= 4\cos\frac{\alpha-\beta}{2}\cos\frac{\gamma-\delta}{2}\left[\cos\frac{\alpha+\beta+\gamma+\delta}{2} + i\sin\frac{\alpha+\beta+\gamma+\delta}{2}\right]$$

(ii) $ab + cd$

$$= 2\cos\frac{\alpha+\beta-\gamma-\delta}{2}\left[\cos\frac{\alpha+\beta+\gamma+\delta}{2} + i\sin\frac{\alpha+\beta+\gamma+\delta}{2}\right]$$

Solution :

(i) $a + b = (\cos\alpha + \cos\beta) + i(\sin\alpha + \sin\beta)$

$$\Rightarrow \; a + b = 2\cos\frac{\alpha+\beta}{2}\cos\frac{\alpha-\beta}{2} + 2i\sin\frac{\alpha+\beta}{2}\cos\frac{\alpha-\beta}{2}$$

$$c + d = 2\cos\frac{\gamma+\delta}{2}\cos\frac{\gamma-\delta}{2} + 2i\sin\frac{\gamma+\delta}{2}\cos\frac{\gamma-\delta}{2}.$$

$$\text{Now } (a+b)(c+d) = 4\cos\frac{\alpha-\beta}{2}\cos\frac{\gamma-\delta}{2}\left\{\cos\frac{\alpha+\beta}{2} + i\sin\frac{\gamma+\beta}{2}\right\}$$

$$\times\left\{\cos\frac{\gamma+\delta}{2} + i\sin\frac{\gamma+\delta}{2}\right\}$$

$$= 4\cos\frac{\alpha-\beta}{2}\cos\frac{\gamma-\delta}{2}\left[\cos\left(\frac{\alpha+\beta}{2}+\frac{\gamma+\delta}{2}\right)\right.$$

$$\left. + i\sin\left(\frac{\alpha+\beta}{2}+\frac{\gamma+\delta}{2}\right)\right].$$

(iii) $ab + cd = \{\cos(\alpha+\beta) + i\sin(\alpha+\beta)\} + \{\cos(\gamma+\delta)$
$+ i\sin(\gamma+\delta)\}$

$$= \{\cos(\alpha+\beta) + \cos(\gamma+\delta)\} + i\sin(\alpha+\beta) + \sin(\gamma+\delta)\}$$

$$= 2\cos\frac{\alpha+\beta+\gamma+\delta}{2}\cos\frac{\alpha+\beta-\gamma-\delta}{2}$$

$$+ 2i\sin\frac{\alpha+\beta+\gamma+\delta}{2}\cos\frac{\alpha+\beta-\gamma-\delta}{2}$$

$$= 2\cos\frac{\alpha+\beta-\gamma-\delta}{2}\left[\cos\frac{\alpha+\beta-\gamma-\delta}{2} + i\sin\frac{\alpha+\beta+\gamma+\delta}{2}\right]$$

Example 43:

If $z = \cos 2\theta + i\sin 2\theta$, $\omega = \cos 2\phi + i\sin 2\phi$; show that

(a) $z^m w^n + \dfrac{1}{z^m\omega^n} = 2\cos 2(m\theta + n\phi)$,

(b) $\dfrac{z^m}{\omega^n} + \dfrac{\omega^n}{z^m} = 2\cos 2(m\theta - n\phi)$, *m and n being integers.*

Solution :

We have $z^m = (\cos 2\theta + i\sin 2\theta)^m$
$= \cos 2m\theta + i\sin 2m\theta$,

and $w^n = (\cos 2\phi + i\sin 2\phi)^n = \cos 2n\phi + i\sin 2n\phi$.

Since $\frac{1}{z} = \cos 2\theta - i \sin 2\theta$, $\frac{1}{z^m} = \cos 2m\theta - i \sin 2m\theta$.

Similarly $\frac{1}{\omega^n} = \cos 2n\phi - i \sin 2n\phi$.

(a) $z^m w^n + \frac{1}{z^m \omega^n} = (\cos 2\ m\theta + i \sin 2m\theta)(\cos 2n\phi + i \sin 2n\phi)$

$+ \{\cos 2n\phi - i \sin 2n\theta\}\{\cos 2n\phi - i \sin 2n\phi\}$

$= [\cos 2(n\theta + n\phi) + i \sin 2(m\theta + n\phi)$

$+ [\cos 2(m\theta + n\phi) - i \sin 2(m\theta + m\phi)]$

$= 2 \cos 2(m\theta + n\phi)$.

(b) $\frac{z^m}{\omega^n} + \frac{\omega^n}{z^m} = (\cos 2m\theta + i \sin 2m\theta)(\cos 2n\phi - i \sin 2n\phi)$

$+ (\cos 2n\phi + i \sin 2n\phi)(\cos 2m\theta - i \sin 2m\theta)$

$= [\cos 2(m\theta - n\phi) + i \sin 2(m\theta - n\phi)]$

$+ [\cos 2(m\theta - n\phi) - i \sin 2(m\theta - n\phi)]$

$= 2 \cos 2(m\theta - n\phi)$.

Example 44:

By expanding $(1 + x)^n$ by Binomial Theorem and then putting $x = \cos 2\theta + i \sin 2\theta + i \sin 2\theta$, prove that

(i) $2^n \cos^n \theta \cos n\theta = \sum_{r=0}^{n} {}^nC_r \cos 2r\theta$

and (ii) $2^n \theta \sin n\theta = \sum_{r=0}^{n} {}^nC_r \sin 2r\theta$.

Solution :

By Binomial Theorem,

$$(1 + x)^n = {}^nC_0 + {}^nC_1 x + {}^nC_2 x^2 + ... + {}^nC_n x^n$$

i.e., $(1 + x)^n \sum_{r=0}^{n} {}^nC_r x^r$.

Putting $x = \cos 2\theta + i \sin 2\theta$, we get

$$(1 + \cos 2\theta + i \sin 2\theta)^n = \sum_{r=0}^{n} {}^nC_r (\cos 2\theta + i \sin 2\theta)^r$$

$$\Rightarrow (2 \cos^2 \theta + i\, 2 \sin \theta \cos \theta)^n = \sum_{r=0}^{n} {}^nC_r (\cos 2r\theta + i \sin 2r\theta)$$

$$\Rightarrow \quad 2^n \cos^n \theta\,[\cos\theta + i\sin\theta]^n = \sum_{r=0}^{n} {}^nC_r \cos 2r\theta + i\sum_{r=0}^{n} {}^nC_r \sin 2r\theta$$

$$\Rightarrow \quad 2^n \cos^n \theta\,(\cos n\theta + i\sin n\theta) = \sum_{r=0}^{n} {}^nC_r \cos 2r\theta + i\sum_{r=0}^{n} {}^nC_r \sin 2r\theta.$$

Equating real and imaginary parts, we get

$$2^n \cos^n \theta \cos n\theta = \sum_{r=0}^{n} {}^nC_r \cos 2r\theta,$$

and $$2^n \cos^n \theta \sin n\theta = \sum_{r=0}^{n} {}^nC_r \sin 2r\theta.$$

Example 45:

If l, m, n,.... are positive integers and

$2\cos\theta = a + 1/a$, $2\cos\phi = b + 1/b$, etc. show that

$$a^l b^m c^n \ldots + \frac{1}{a^l b^m c^n \ldots} = 2\cos(l\theta + m\phi + \ldots).$$

Solution :

$$2\cos\theta = a + \frac{1}{a} \Rightarrow a^2 - 2a\cos\theta + 1 = 0.$$

$$\therefore \quad a = \frac{2\cos\theta + \sqrt{4\cos^2\theta - 4}}{2} = \cos\theta \pm i\sin\theta.$$

We may take $a = \cos\theta + i\sin\theta$.

Now $a^l = (\cos\theta + i\sin\theta)^l = \cos l\theta + i\sin l\theta$.

Similarly $b^m = (\cos\phi + i\sin\phi) = \text{cow } m\phi + i\sin m\phi$ etc.

Thus $a^l b^m c^n .... = (\cos l\theta + i\sin l\theta)(\cos m\phi + i\sin m\phi).....$

$\Rightarrow \quad a^l b^m c^n .... = \cos(l\theta + m\phi + \ldots) + i\sin(l\theta + m\phi + \ldots).$

$$\therefore \quad \frac{1}{a^l b^m c^n \ldots} = \cos(l\theta + m\phi + \ldots) - i\sin(l\theta + m\phi + ...).$$

Hence $a^l b^m c^n \ldots + \dfrac{1}{a^l b^m c^n \ldots} = 2\cos(l\theta + m\phi + ...).$

POLAR FORM

Let $z = a + ib$. On putting $a = r\cos\theta$, $b = r\sin\theta$; we get

$$z = r(\cos\theta + i\sin\theta),$$

$\Rightarrow \quad r = \sqrt{a^2 + b^2}$ and $\tan\theta = b/a$.

Now $z = r(\cos\theta + i\sin\theta)$ and by De Moivre's Theorem,

$$z^n = r^n(\cos n\theta + i\sin n\theta), = r = \sqrt{a^2+b^2}, \theta = \tan^{-1}(b/a).$$

Example 1:

If n is a positive integer, show that

$$\left(\sqrt{3+i}\right)^n + \left(\sqrt{3-i}\right)^n = 2^{n+1}\cos(n\pi/6).$$

Solution :

Let $\sqrt{3} = r\cos\theta$ and $1 = r\sin\theta$ so that

$$r^2 = 4 \text{ and } \tan\theta = 1/\sqrt{3} \Rightarrow r = 2, \theta = \pi/6.$$

$$\therefore \left(\sqrt{3+i}\right)^n = 2^n(\cos\pi/6 + i\sin\pi/6)^n$$

$$\Rightarrow \left(\sqrt{3+i}\right)^n = 2^n\{\cos(n\pi/6) + i\sin(n\pi/6)\}. \quad ...(1)$$

Similarly $\left(\sqrt{3-i}\right)^n = 2^n\{\cos(n\pi/6) - i\sin(n\pi/6)\}. \quad ...(2)$

Adding (1) and (2), we obtain

$$\left(\sqrt{3+i}\right)^n + \left(\sqrt{3-i}\right)^n = 2.2^n\cos(n\pi/6) = 2^{n+1}\cos(n\pi/6).$$

Example 2:

If n be a positive integer, prove that

$$\left(1+i\sqrt{3}\right)^n + \left(1-i\sqrt{3}\right)^n = 2^{n-1}\cos\frac{n\pi}{3}.$$

Solution :

Let $r\cos\theta = 1$ and $r\sin\theta = \sqrt{3}$.

$$\Rightarrow r^2 = 4 \text{ and } \tan\theta = \sqrt{3} \Rightarrow r = 2, \theta = \pi/3.$$

$$\therefore 1 \pm i\sqrt{3} = 2\left(\cos\frac{\pi}{3} \pm i\sin\frac{\pi}{3}\right).$$

Now $\left(1+i\sqrt{3}\right)^n + \left(1-i\sqrt{3}\right)^n$

$$= 2^n(\cos\pi/3 + i\sin\pi/3)^n + 2^n(\cos\pi/3 - i\sin\pi/3)^n$$

$$= 2^n\left(\cos\frac{n\pi}{3} + i\sin\frac{n\pi}{3}\right) + 2^n\left(\cos\frac{n\pi}{3} - i\sin\frac{n\pi}{3}\right)$$

$$= 2^n \left(2 \cos \frac{n\pi}{3}\right) = 2^{n+1} \cos \frac{n\pi}{3}.$$

Example 3:

Prove that $(1 + i)^n + (1 - i)^n = 2^{\frac{n}{2}+1} \cos (n\pi/4)$.

Solution :

$1 = r \cos \theta$ and $1 = r \sin \theta \Rightarrow r^2 = 2$ and $\tan \theta = 1$.

$\Rightarrow$ $r = \sqrt{2}$ and $\theta = \pi/4$, and so

$$1 + i = \sqrt{2} (\cos \pi/4 + i \sin \pi/4)$$

$\Rightarrow$ $(1 + i)^n = 2^{n/2} (\cos \pi/4 + i \sin \pi/4)^n = 2^{n/2} [\cos n\pi/4 + i \sin n\pi/4]$ ...(1)

Similarly $(1 - i)^n = 2^{n/2} [\cos n\pi/4 - i \sin n\pi/4]$. ...(2)

Adding (1) and (2), we obtain

$$(1 + i)^n + (1 - i)^n = 2^{n/2} \cos (n\pi/4) = 2^{\frac{n}{2}+1} \cos (n\pi/4).$$

Example 4:

If α, β, *be the rots of* $x^2 - 2x + 2 = 0$, *prove that*

$$\alpha^n + \beta^n = 2^{\frac{n}{2}+1} \cos (n\pi/4).$$

Solution :

Solving $x^2 - 2x + 2 = 0$, we obtain

$$x = \frac{2 \pm \sqrt{4-8}}{2} = 1 + i.$$

Thus $a = 1 + i$ and $b = 1 - i$ and so

$$\alpha^n + \beta^n = (1 + i)^n + (1 - i)^n$$

$$= 2^{\frac{n}{2}+1} \cos (n\pi/4).$$

Example 5:

If $(1 + z)^n = c_0 + c_1 z + c_2 z^2 + ... + c_n z^n$ *where n is is a positive integer, show that*

$$c_0 - c_2 + c_4 - ... = 2^{n/2} \cos (n\pi/4),$$

$$c_1 - c_3 + c_5 - ... = 2^{n/2} \sin (n\pi/4).$$

Solution :

We have

$(1 + z)^n = c_0 + c_1z + c_2z^2 + c_3z^3 + c_4z^4 + ... + c_nz^n.$

Putting z = i, we obtain

$(1 + i)^n = c_0 + c_1i + c_2i^2 + c_3i^3 + c_4i^4 + c_5i^5 + ...$

$= c_0 + c_1i - c_2 - c_3i + c_4 + c_5i.....$

$\therefore \quad (1 + i)^n = (c_0 - c_2 + c_4 - ....) + i\,(c_1 - c_2 + c_5 - .....)$....(1)

Let $1 + r\cos\theta$ and $1 = r\sin\theta$ so that

$$r^2 = 2 \text{ and } \tan\theta = 1 \Rightarrow r = \sqrt{2} \text{ and } \theta = \pi/4.$$

$$\therefore \quad 1 + i = r(\cos\theta + i\sin\theta) = \sqrt{2}\,(\cos\pi/4 + i\sin\pi/4)$$

$$\Rightarrow (1 + i)^n = 2^{n/2}(\cos\pi/4 + i\sin\pi/4)^n$$

$$\Rightarrow (1 + i)^n = 2^{n/2}\left[\cos\frac{n\pi}{4} + i\sin\frac{n\pi}{4}\right]. \qquad ...(2)$$

Equating real and imaginary parts on the R.H.S. of (1) and (2), we obtain

$c_0 - c_2 + c_4 - ..... = 2^{n/2}\cos(n\pi/4),$

and $c_1 - c_3 + c_5 - ..... = 2^{n/2}\sin(n\pi/4).$

Example 6:

If $(a_1 + ib_1)(a_2 + ib_2).....(a_n + ib_n) = A + iB$, prove that

(i) $\left(a_1^2 + b_1^2\right)\left(a_2^2 + b_2^2\right).....\left(a_n^2 + b_n^2\right) = A^2 + B^2$

(ii) $\tan^{-1}\dfrac{b_1}{a_1} + \tan^{-1}\dfrac{b_2}{a_2} + ...+ \tan^{-1}\dfrac{b_n}{a_n} = \tan^{-1}\dfrac{B}{A}.$

Solution :

Let $a_1 + ib_1 = r_1(\cos\theta_1 + i\sin\theta_1)$, so that

$a_1 = r_1\cos b_1$ and $\theta_1 = r_1\sin\theta_1$.

$\Rightarrow \quad r_1^2 = a_1^2 + b_1^2$ and $\tan\theta_1 = b_1/a_1 \Rightarrow \theta_1 = \tan^{-1} b_1/a_1$.

Similarly $a_2 + ib_2 = r_2(\cos\theta_2 + i\sin\theta_2)$ gives

$r_2^2 = a_2^2 + b_2^2$ and $\theta_2 = \tan^{-1} b_2/a_2$ and so on.

Consider $(a_1 + ib_1)(a_2 + ib_2)....(a_n + ib_n) = A + iB$.

$\Rightarrow \quad r_1(\cos\theta_1 + i\sin\theta_1).r_2(\cos\theta_2 + i\sin\theta_2)...r_n(\cos\theta_n + i\sin\theta_n)$

$= A + iB.$

$\Rightarrow$ $r_1 r_2 .....r_n$ $(\cos\theta_1 + i \sin\theta_1)(\cos\theta_2 + i \sin\theta_2) + i \sin\theta_2)$
$...(\cos\theta_n + i \sin\theta_n)$

$= A + iB.$

$\Rightarrow$ $r_1 r_2 ....r_n[\cos(\theta_1 + \theta_2 + ... + \theta_n) + i \sin(\theta_1 + \theta_2 + ... + \theta_n)]$
$= A + iB.$

Comparing real and imaginary parts on both sides, we get

$r_1 r_2 ....r_n \cos(\theta_1 + \theta_2 + ... + \theta_n) = A,$ ...(1)

$r_1 r_2 .....r_n \sin(\theta_1 + \theta_2 + ... + \theta_n) = B,$ ...(2)

Squaring and adding (1) and (2), we get

$$r_1^2 r_2^2 .....r_n^2 = A^2 + B^2.$$

Hence $\left(a_1^2 + b_1^2\right)\left(a_2^2 + b_2^2\right).......\left(a_n^2 + b_n^2\right) = A^2 + B^2.$

Dividing (2) by (1), we get

$\tan(\theta_1 + \theta_2 + ... + \theta_n) = B/A$

$\Rightarrow$ $\theta_1 + \theta_2 + ... + \theta_n = \tan^{-1}(B/A).$

Hence $\tan^{-1}(b_1/a_1) + \tan^{-1}(b_2/a_2) + ... + \tan^{-1}(b_n/a_n)$
$= \tan^{-1}(B/A).$

Example 7:

If the roots of $t^2 - 2t + 2 = 0$ are α and β, s how that

(a) $$\frac{(x+\alpha)^n + (x+\beta)^n}{\alpha+\beta} = \frac{\cos n\theta}{\sin^n \phi}$$

(b) $$\frac{(x+\alpha)^n + (x+\beta)^n}{\alpha+\beta} = \frac{\cos n\theta}{\sin^n \phi},$$ *where $x + 1 = \cot\phi$.*

Solution :

Solving $t^2 - 2t + 2$, we get

$$\Rightarrow \quad t = \frac{2 \pm \sqrt{4-8}}{2} = \frac{2 \pm 2i}{2} = 1 \pm i.$$

Let $\alpha = 1 + i$ and $\beta = 1 - i$.

$\therefore$ $x + \alpha = (x + 1) + i = r(\cos\phi + i \sin\phi),$

and $x + \beta = (x + 1) - i = r(\cos\phi - i \sin\phi),$

where $r \cos\phi = x + 1$, $r \sin\phi = 1$.

$\therefore$ $\cot\phi = (x + 1)$ and $r = \sqrt{1+(x+1)^2} = \sqrt{1+\cot^2\phi} =$ coses ϕ.

(a) $\dfrac{(x+\alpha)^n-(x+\beta)^n}{\alpha-\beta}$

$= \dfrac{r^n\,(\cos n\phi + i \sin n\phi) - r^n\,(\cos n\phi - i \sin n\phi)}{\alpha-\beta}$

$= \dfrac{r^n\,(2i \sin n\phi)}{1+i-1+i} = r^n \;\sin n\phi$

$= \dfrac{\sin n\phi}{\sin^n \phi}.\quad (\because r \sin \phi = 1)$

(b) $\dfrac{(x+\alpha)^n+(x+\beta)^n}{\alpha+\beta}$

$= \dfrac{r^n\,(\cos n\phi + i \sin n\phi) + r^n\,(\cos n\phi - i \sin n\phi)}{1+i+1-i}$

$= r^n \cos nf = \dfrac{\cos n\phi}{\sin^n \phi}.$

Example 8:

If $\sin\left(\frac{1}{6}\pi+\theta\right) = 0.51$, *prove that* θ *is equal to 40′ nearly.*

Solution :

$\because$ 0.51 is nearly equal to $\frac{1}{2}$ *i.e.,* $\sin\frac{1}{6}\pi$

$\therefore$ θ must be small.

Now the given equation is $\sin\left(\frac{1}{6}\pi+\theta\right) = 0.51$

$\Rightarrow \sin\frac{1}{6}\pi \;\cos\theta + \cos\frac{1}{6}\pi\;\sin\theta = 0.51$

$\Rightarrow \frac{1}{2}\cos\theta + \left(\sqrt{3/2}\right)\sin\theta = 0.51$

$\Rightarrow \frac{1}{2}\left(1-\frac{\theta^2}{2\,!}+\ldots\right)+\frac{\sqrt{3}}{2}\left(\theta-\frac{\theta^3}{3\,!}+\ldots\right) = 0.51$

$\frac{1}{2}\left(1-\frac{1}{2}\theta^2\right)+\frac{1}{2}\sqrt{3}\,(\theta) = 0.51$, neglecting powers of θ higher than the second.

$\Rightarrow -\frac{1}{4}\theta^2+\frac{1}{2}\sqrt{3q} = .51-\frac{1}{2} = 0.51-0.50 = 0.01 = \frac{1}{100}$

$$\Rightarrow \; 25\theta^2 - 50\sqrt{3\theta} + 1 = 0$$

$$\Rightarrow \; \theta = 0.012 \text{ radian} = \frac{12}{1000} \times \frac{180 \times 7}{22} \times 60 \text{ min} = 40' \text{ nearly.}$$

Example 9:

If $\cos\left(\frac{1}{3}\pi + \theta\right) = 0.49$*, prove that θ is equal to 40′ nearly.*

Solution :

0.49 is nearly equal to $\frac{1}{2}$ *i.e.,* $\cos \frac{1}{2}\pi$

$\therefore$ θ must be very small.

Now the given equation is $\cos\left(\frac{1}{3}\pi + \theta\right) = 0.49$

$$\Rightarrow \; \cos\left(\frac{1}{2}\pi\right) \cos\theta + \sin\left(\frac{1}{2}\pi\right) \sin\theta = 0.49$$

$$\Rightarrow \; \frac{1}{2}\cos\theta - \left(\sqrt{3/2}\right)\sin\theta = 0.49$$

$$\Rightarrow \; \frac{1}{2}\left(1 - \frac{\theta^2}{2!} + \ldots\right) - \frac{\sqrt{3}}{2}\left(\theta - \frac{\theta^3}{3!} + \ldots\right) = 0.49$$

$$\Rightarrow \; \frac{1}{2}\left(1 - \frac{\theta^2}{2!}\right) - \frac{\sqrt{3}}{2}(\theta) = 0.49, \text{ neglecting powers of } \theta \text{ higher than}$$

the second.

$$\frac{1}{2} - \frac{\theta^2}{4} - \frac{\theta\sqrt{3}}{2} = C.49 = \frac{49}{100} \Rightarrow 25\theta^2 + 50\sqrt{3\theta} - 1 = 0$$

$$\Rightarrow \; \theta = 0.12 \text{ radian nearly} = \frac{12}{1000} \times \frac{180}{22} \times 7 \times 60 \text{ min.}$$

$$= 40 \text{ minutes nearly.}$$

Hence proved.

3

FRACTIONAL INDEX

DE MOIVRE'S THEOREM FOR A FRACTIONAL INDEX

Theorem 1:

State and prove De Moivre's Theorem for a fractional index.

Or

If p and q be integers ($q \neq 0$), then cos (p^{θ}/q) + i sin (p^{θ}/q) is one of the values of ($\cos \theta + i \sin \theta)^{p/q}$.

Proof:

By De Moivre's Theorem for an integral index,

$$\left(\cos \frac{\theta}{q} + i \sin \frac{\theta}{q}\right)^q = \cos q.\frac{\theta}{q} + i \sin q.\frac{\theta}{q}$$

$$i = \cos \theta + i \sin \theta.$$

So the qth power of $\cos \theta/q + i \sin \theta/q$ is $\cos \theta + i \sin \theta$,

i.e., ($\cos \theta/q + i \sin \theta/q$) is one of the qth roots of $\cos \theta + i \sin\theta$,

i.e., ($\cos \theta/q + i \sin \theta/q$) is one of the values of $(\cos \theta + i \sin \theta)^{1/q}$.

Raising both sides to the power p, we see that

$(\cos \theta/q + i \sin \theta/q)^p$ is one of the values of $(\cos \theta + i \sin \theta)^{p/q}$.

Hence $\cos p^{\theta}/q + i \sin p^{\theta}/q$ is one of the values of

$(\cos \theta + i \sin \theta)^{p/q}$. *(De Moivre's Theorem for an integral index)*

Theorem 2:

If p and q be integers prime to each other ($\theta \times 0$), then all the values of $(\cos \theta + i \sin \theta)^{p/q}$ are given by

$$\cos\left[(2\,k\pi + \theta)\,\frac{p}{q} + i\sin(2k\pi + \theta)\,\frac{p}{q}\right],$$

where k = 0, 1, 2,...., q – 1.

The proof is omitted.

Proof:

$$\left(\cos\frac{\pi}{3} + i\sin\frac{\pi}{3}\right)^{3/5}$$

$$= \cos\left(2k\pi + \frac{\pi}{3}\right).\frac{3}{5} + i\sin\left(2k\pi + \frac{\pi}{3}\right).\frac{3}{5},$$

where k = 0, 1, 2, 3, 4.

Example 1:

Write down all the values of $(i)^{1/6}$.

Solution :

We have i = 0 + i.1 = π/2 + i sin π/2. Thus

$(i)^{1/6} = (\cos\pi/2 + i\sin\pi/2)^{1/6}$.

By Theorem 3,

$$(i)^{1/6} = \left[\cos\left(2k\pi + \frac{\pi}{2}\right)\frac{1}{6} + i\sin\left(2k\pi + \frac{\pi}{2}\right)\frac{1}{6}\right],$$

where k = 0, 1, 2, 3, 4, 5

$$= \cos(4k + 1)\,\frac{\pi}{12} + i\sin(4k + 1)\,\frac{\pi}{12}$$

$$= \cos\left(\frac{p\pi}{12}\right) + i\sin\left(\frac{p\pi}{12}\right),\ p = 1, 5, 9, 13, 17, 21.$$

Example 2:

Write down all the values of

(a) $\left(1+i\sqrt{3}\right)^{2/3}$, *(b)* $\left(\sqrt{3}-i\right)^{3/5}$, *(c)* $(1 + i)^{1/3}$.

Solution :

(a) Let $1 + i\sqrt{3} = r(\cos\theta + i\sin\theta)$.

Then $1 = r\cos\theta$ and $\sqrt{3} = r\sin\theta$.

$\Rightarrow r^2 = 4$ and $\tan\theta = \sqrt{3} \Rightarrow r = 2,\ \theta = \pi/3$.

Now $\left(1+\sqrt{3}\right)^{2/3} = r^{2/3}\ (\cos\theta + i\sin\theta)^{2/3}$

$= 2^{2/3}\ (\cos\pi/3 + i\sin\pi/3)^{2/3}$

$$= 2^{2/3}\left[\cos\frac{2}{3}\left(2k\pi + \frac{\pi}{3}\right) + i\sin\frac{2}{3}\left(2k\pi + \frac{\pi}{3}\right)\right],\ k = 0, 1, 2.$$

$$= 2^{2/3}\left[\cos\frac{2(6k+1)\pi}{9} + i\sin\frac{2(6k+1)\pi}{9}\right],\ k = 0, 1, 2.$$

Hence $\left(1+i\sqrt{3}\right)^{2/3} = 2^{2/3}\left[\cos\frac{p\pi}{9} + i\sin\frac{p\pi}{9}\right],\ p = 2, 14, 26.$

(b) Let $\sqrt{3}-1 = r\ (\cos\theta - i\sin\theta)$.

Then $\sqrt{3} = r\cos\theta,\ 1 = r\sin\theta$

$\Rightarrow\ r^2 = 4$ and $\tan\theta = 1/\sqrt{3} \Rightarrow r = 2$ and $\theta = \pi/6$.

Now $\left(\sqrt{3}-i\right)^{3/5} = (2)^{3/5}\ (\cos\pi/6 - i\sin\pi/6)^{3/5}$

$= (2)^{3/5}\ (\cos\pi/2 - i\sin\pi/2)^{1/5}$. *(De Moivre's Theorem)*

$$= (2)^{3/5}\left[\cos\frac{3\pi}{2} + i\sin\frac{3\pi}{2}\right]^{1/5}$$

$$= (2)^{3/5}\left[\cos\left(2k\pi + \frac{3\pi}{2}\right)\frac{1}{5} + i\sin\left(2k\pi + \frac{3\pi}{2}\right)\frac{1}{5}\right],$$

$k = 0, 1, 2, 3, 4$

$$= (2)^{3/5}\left[\cos(4k+3)\frac{\pi}{10} + i\sin(4k+3)\frac{\pi}{10}\right],\ k = 0, 1, 2, 3, 4$$

Hence $\left(\sqrt{3}-i\right)^{3/5} = (2)^{3/5}\ \cos\frac{p\pi}{10} + i\sin\frac{p\pi}{10},\ p = 3, 7, 11, 15, 19.$

(c) Let $1 + i = r\ (\cos\theta + i\sin\theta)$.

Then $1 = r\cos\theta$ and $1 = r\sin\theta$.

$\Rightarrow$ $r^2 = 2$ and $\tan\theta = 1 \Rightarrow r = \sqrt{2}$ and $\theta = \pi/4$.

Now $(1 + i)^{1/3} = \left(\sqrt{2}\right)^{1/3}\ (\cos\pi/4 + i\sin\pi/4)^{1/3}$.

$$= \left(2^{1/2}\right)^{1/3}\left[\cos\frac{1}{3}\left(2k\pi + \frac{\pi}{4}\right) + i\sin\frac{1}{3}\left(2k\pi + \frac{\pi}{4}\right)\right],\ k = 0, 1, 2,$$

$$= 2^{1/6}\left[\cos(8k+1)\frac{\pi}{12} + i\sin(8k+1)\frac{\pi}{12}\right],\ k = 0, 1, 2.$$

Hence $(1 + i)^{1/3} = 2^{1/6} \left[\cos\left(\frac{p\pi}{12}\right) + i \sin\left(\frac{p\pi}{12}\right)\right]$, $\pi = 1, 9, 17.$

Example 3:

Find all the values of $\left(\frac{1}{2} + i\frac{\sqrt{3}}{2}\right)^{3/4}$ *and show that the continued product of all the values is 1.*

Solution :

Let $\left(\frac{1}{2} + i\frac{\sqrt{3}i}{2}\right) = r(\cos\theta + i\sin\theta).$

$\therefore$ $1/2 = r\cos\theta$ and $\sqrt{3/2} = r\sin\theta,$

$\Rightarrow$ $r^2 = 1$ and $\tan\theta = \sqrt{3} \Rightarrow r = 1$ and $\theta = \pi/3.$

Now $\left(\frac{1}{2} + \frac{\sqrt{3}}{2}\right)^{3/4} = \left[(\cos\pi/3 + i\sin\pi/3)^3\right]^{1/4}$

$= (\cos\pi + i\sin\pi)^{1/4}$

$= [\cos(2k\pi + \pi) + i\sin(2k\pi + \pi)^{1/4}, k = 0, 1, 2, 3$

$= \cos(2k+1)\frac{\pi}{4} + i\sin(2k+1)\frac{\pi}{4}, k = 0, 1, 2, 3.$

Hence the required four values are, respectively

$\cos(\pi/4) + i\sin(\pi/4)$, $\cos(3\pi/4) + i\sin(3\pi/4)$,

$\cos(5\pi/4) + i\sin(5\pi/4)$ and $\cos(7\pi/4) + i\sin(7\pi/4)$.

The combined product of the above four values

$$= \cos\left(\frac{\pi}{4} + \frac{3\pi}{4} + \frac{7\pi}{4}\right) + \sin\left(\frac{\pi}{4} + \frac{3\pi}{4} + \frac{5\pi}{4} + \frac{7\pi}{4}\right)$$

$= \cos 4\pi + i\sin 4\pi = 1 + i\,.\,0 = 1.$

Example 4:

Prove that n nth roots of unity form a series in G.P.

Solution :

Let $x = (1)^{1/n} = (\cos 0 + i\sin 0)^{1/n}$

$$= \cos\left(\frac{0 + 2k\pi}{n}\right) + i\sin\left(\frac{0 + 2k\pi}{n}\right) \quad k = 0, 1, 2, \ldots n - 1$$

Thus n roots of unity are

$$1, \cos\left(\frac{2\pi}{n}\right) + i \sin\left(\frac{2\pi}{n}\right), \cos\left(\frac{4\pi}{n}\right) + i \sin\left(\frac{4\pi}{n}\right), \ldots\ldots,$$

$$\cos\left\{\frac{2}{n}(n-1)\pi\right\} + i \sin\left\{\frac{2}{n}(n-1)\pi\right\}$$

$$\Rightarrow \quad 1, \alpha, \alpha^2, \ldots, \alpha^{n-3}; \text{ where } \alpha = \cos\left(\frac{2\pi}{n}\right) + 1 \sin\left(\frac{2\pi}{n}\right)$$

The above roots $1, \alpha, \alpha^2, \ldots \alpha^{n-1}$ are clearly G.P.

Example 5:

Find the fifth roots of (–32).

Solution :

We have $(-32)^{1/5} = \{2^5 (-1)\}^{1/5} = 2 (-1)^{1/5}$

$= 2 (\cos \pi + i \sin \pi)^{1/5}$

$$= 2\left[\cos \frac{1}{5}(2k\pi + \pi) + i \sin \frac{1}{5}(2k\pi + \pi)\right], k = 0, 1, 2, 3, 4.$$

Hence $(-32)^{1/5} = 2 \{\cos (p\pi/5) + i \sin (p\pi/5)\}$, $p = 1, 3, 5, 7, 9$.

Example 6:

We have $(16)^{1/4} = (2^4)^{1/4} = 2 (1)^{1/4}$

$= 2 (\cos 0 + i \sin 0)^{1/4}$

$$= 2 \{\cos \frac{1}{4} (2k\pi + 0) + i \sin \frac{1}{4} (2k\pi + 0)\}, k = 0, 1, 2, 3$$

$= 2 \times 1, 2 \times i, 2 \times -1, 2 \times -i = \pm 2, \pm 2i.$

Example 7:

Show that one of the values of

$$(a + ib)^{m n} + (a - ib)^{m n} \text{ is } 2 (a^2 + b^2)^{m 2n} \cos\left(\frac{m}{n} \tan^{-1} \frac{b}{a}\right).$$

Solution :

Let $a + ib = r (\cos \theta + i \sin \theta)$. ...(i)

Then $r \cos \theta = a$ and $r \sin \theta = b$.

$\Rightarrow$ $r^2 = a^2 + b^2$, $\tan \theta = b/a$

$\Rightarrow$ $\theta = ta^{-1}\, b/a$. ...(ii)

Taking conjugates of both sides of (i), we get

$a - ib = r(\cos\theta - i\sin\theta)$.

Now by De Moivre's Theorem, one of the values of

$$(a + ib)^{m/n} \text{ is } r^{m/n}\left(\cos\frac{m\theta}{n} + i\sin\frac{m\theta}{n}\right),$$

and one of the values of

$$(a - ib)^{m/n} \text{ is } r^{m/n}\left(\cos\frac{m\theta}{n} - i\sin\frac{m\theta}{n}\right).$$

Hence one of the values of $(a + ib)^{m/n} + (a - ib)^{m/n}$ is

$$= 2r^{m/n}\cos\frac{m\theta}{n}$$

$$= 2(a^2 + b^2)^{m/2n}\cos\left(\frac{m}{n}\tan^{-1}\frac{b}{a}\right), \text{ using (ii).}$$

Example 8:

Write down all the values of each of the following :

(i) $i^{1/3}$ *(ii)* $(1 + i)^{1/4}$ *(iii)* $\left(\sqrt{3+i}\right)^{1/3}$.

Ans.

(i) $\cos(\pi/12) + i\sin(\pi/12)$, $\pi = 1, 5, 9, 13, 17, 21$.

(ii) $2^{1/8}\{\cos(p\pi/16) + i\sin(p\pi/16)\}$, $p = 1, 9, 17, 25$.

(ii) $2^{1/3}\{\cos(p\pi/18) + i\sin(p\pi/18)\}$, $\pi = 1, 13, 25$.

SOLUTION OF EQUATIONS

In this section we shall solve certain equations with the help of De Moivre's Theorem.

Example 1:

Solves the equations :

(i) $x^3 - 1 = 0$ *(ii)* $x^3 + 1 = 0$.

Solution :

(i) $x^3 - 1 = 0 \Rightarrow x^3 = 1 = \cos 0 + i\sin 0$

$\Rightarrow x = \{\cos(2k\pi + 0) + i\sin(2k\pi + 0)\}^{1/8}$

$\Rightarrow x = \cos\frac{2}{3}k\pi + i\sin\frac{2}{3}k\pi$, $k = 0, 1, 2$.

Hence $1, -\frac{1}{2} + i\frac{\sqrt{3}}{2}, -\frac{1}{2} - i\frac{\sqrt{3}}{2}$ are the roots.

(ii) $x^3 + 1 = 0 \Rightarrow x^2 = -1 \Rightarrow x = (\cos \pi + i \sin \pi)^{1/3}$

$$\Rightarrow x = \cos \frac{1}{2}(2k\pi + \pi) + i \sin \frac{1}{2}(2k\pi + \pi),\ k = 0, 1, 2.$$

Hence $\frac{1}{2} + i\frac{\sqrt{3}}{2}, -1, \frac{1}{2} - i\frac{\sqrt{3}}{2}$ are the roots.

Example 2:

Solve the equation $z^4 + 1 = 0$.

Solution :

$z^4 + 1 = 0 \Rightarrow = -1$

$\Rightarrow z = (\cos \pi + i \sin \pi)^{1/4}$

$$\Rightarrow z = \cos \frac{1}{4}(2k\pi + \pi) + i \sin \frac{1}{4}(2k\pi + \pi),\ k = 0, 1, 2, 3.$$

$\Rightarrow z = \cos \pi/4 + i \sin \pi/4, \cos (3\pi/4) + i \sin (3\pi/4),$

$\cos (5\pi/4) + i \sin (5\pi/4), \cos (7\pi/4) + i \sin (7\pi/4)$

$$\frac{1}{\sqrt{2}}(1 + i), \frac{1}{\sqrt{2}}(-1 + i), \frac{1}{\sqrt{2}}(-1 - i), \frac{1}{\sqrt{2}}(1 - i).$$

Hence the four roots of $z^4 + 1 = 0$ are $\frac{1}{\sqrt{2}}(\pm 1 \pm i)$.

Example 3:

Solve the equation $z^5 + 1 = 0$.

Solution :

We have $z^5 + 1 = 0$

$\Rightarrow z^5 = -1 = \cos \pi + i \sin \pi.$

Therefore $z = (\cos \pi + i \sin \pi)^{1/5}$

$$= \cos \frac{1}{5}(2k + 1)\pi + i \sin \frac{1}{5}(2k + 1)\pi,\ k = 0, 1, 2, 3, 4.$$

Hence $\cos (\pi/5) + i \sin (\pi/5), \cos (3\pi/5) + i \sin (3\pi/5), -1,$

$\cos (7\pi/5) + i \sin (7\pi/5), \cos (9\pi/5) + i \sin (9\pi/5)$

are the required roots.

Example 4:

Solve the equation $z^5 - 1 = 0$.

Solution :

$z^5 - 1 = 0 \Rightarrow z^5 = 1 = \cos 0 + i \sin 0$

$\Rightarrow$ $z = (\cos 0 + i \sin 0)^{1/5} = (\cos 2k\pi + i \sin 2k\pi)^{1/5}$

Hence $z = \cos\left(\frac{2}{5}k\pi\right) + i \sin\left(\frac{2}{5}k\pi\right)$, $k = 0, 1, 2, 3, 4.$

Example 5:

Solve the equation $x^6 + 1 = 0$.

Solution :

We have $x^6 = -1 = \cos \pi + i \sin \pi$

$\Rightarrow$ $x = [\cos (2k\pi + \pi) + i \sin (2k\pi + 1)^{1/6}$

$\Rightarrow$ $x = [\cos (2k + \pi)\frac{\pi}{6} + i \sin (2k + 1)\frac{\pi}{6}$, $k = 0, 1, 2, 3, 4, 5.$

$= \cos \frac{p\pi}{6} + i \sin \frac{p\pi}{6}$, $p = 1, 3, 5, 7, 9, 11$

$= \cos \pi/6 + i \sin \pi/6, i, -\cos \pi/6 + i \sin \pi/6,$

$-\{\cos \pi/6 + i \sin \pi/6\}, -i, \cos \pi/6 - i \sin \pi/6.$

$[\because 5\pi/6 = \pi - \pi/6, 7\pi/6 = \pi + \pi/6]$

Hence the six values are

$$\pm i, \frac{\sqrt{3}}{2} \pm \frac{i}{2}, -\frac{\sqrt{3}}{2} \pm \frac{i}{2}.$$

Example 6:

Solve the equation

(i) $z^7 - z = 0$ *(ii) $z^5 - z = 0$.*

Solution :

(i) $z^7 - z = 0 \Rightarrow z(z^6 - 1) = 0$

$\Rightarrow z = 0 \Rightarrow z^6 = 1.$

Now $z^6 = 1 = \cos 0 + i \sin 0$

$\therefore$ $z = \cos\left(\frac{0 + 2k\pi}{6}\right) + i \sin\left(\frac{0 + 2k\pi}{6}\right)$, $k = 0, 1, 2, 3, 4, 5.$

$= \cos\left(\frac{k\pi}{3}\right) + i \sin\left(\frac{k\pi}{3}\right)$, $k = 0, 1, 2, 3, 4, 5.$

Hence $1, \frac{1}{2} + i\frac{\sqrt{3}}{2}, -\frac{1}{2} + i\frac{\sqrt{3}}{2}, -1, -\frac{1}{2} - i\frac{\sqrt{3}}{2}$

$\frac{1}{2} - i\frac{\sqrt{3}}{2}$ and 0 are the required seven roots of $z^7 - z = 0$.

(ii) $z^5 - z = z(z^4 - 1) = z(z^2 - 1)(z^2 + 1)$.

Hence the roots of $z^5 - z = 0$ are $0, \pm 1, \pm i$.

Example 7:

Solve the equation $x^7 - 1 = 0$.

Solution :

$$x^7 - 1 = 0 \Rightarrow x = (1)^{1/7} = (\cos 0 + i \sin 0)^{1/7}$$

$$\therefore \quad x = \cos \frac{2k\pi}{7} + i \sin \frac{2k\pi}{7}, \ k = 0, 1, 2, 3, 4, 5, 6.$$

Thus the seven values are

$\cos 0 + i \sin 0$, $\cos (2\pi/7) + i \sin (2\pi/7)$, $\cos (4\pi/7) + i \sin (4\pi/7)$,

$\cos (6\pi/7) + i \sin (6\pi/7)$, $\cos (8\pi/7) + i \sin (8\pi/7)$,

$\cos (10\pi/7) + i \sin (10\pi/7)$, $\cos (12\pi/7) + i \sin (12\pi/7)$.

$$\text{Now } \cos \frac{8\pi}{7} + i \sin \frac{8\pi}{7} = \cos\left(2\pi - \frac{6\pi}{7}\right) + i \sin\left(2\pi - \frac{6\pi}{7}\right)$$

$$= \cos \frac{6\pi}{7} - i \sin \frac{6\pi}{7}.$$

$$\text{Similarly } \cos \frac{10\pi}{7} + i \sin \frac{10\pi}{7} = \cos \frac{4\pi}{7} - i \sin \frac{4\pi}{7}.$$

$$\text{and } \cos \frac{12\pi}{7} + i \sin \frac{12\pi}{7} = \cos \frac{2\pi}{7} - i \sin \frac{2\pi}{7}.$$

Hence the seven values are

$$1, \left(\cos \frac{2\pi}{7} \pm i \sin \frac{2\pi}{7}\right), \left(\cos \frac{4\pi}{7} \pm i \sin \frac{4\pi}{7}\right)$$

$$\text{and } \left(\cos \frac{6\pi}{7} \pm i \sin \frac{6\pi}{7}\right).$$

Example 8:

Solve the equations :

(a) $x^7 - x^4 + x^3 - 1 = 0.$

(b) $x^6 - x^5 + x^4 - x^3 + x^2 - x + 1 = 0.$

Solution :

(a) We see that

$$x^7 - x^4 + x^3 - 1 = 0 \Rightarrow x^4 (x^3 - 1) + (x^3 - 1) = 0$$

$$\Rightarrow (x^4 + 1)(x^3 - 1) = 0 \Rightarrow x^3 - 1 = 0 \Rightarrow x^4 + 1 = 0.$$

$1, \frac{1}{2}\left(-1 \pm i\sqrt{3}\right), \frac{1}{\sqrt{2}}\ (\pm 1 \pm i).$

(b) We have $x^6 - x^5 + x^4 - x^3 + x^2 - x + 1 = 0.$

Multiplying the given equation by x + 1, we get

$$x^7 + 1 = 0 \Rightarrow x^7 = -1 = \cos \pi + i \sin \pi.$$

$$\Rightarrow x = [\cos (2k\pi + \pi) + i \sin (2k\pi + \pi)]^{1/7}$$

$$= \cos (2k\pi + 1)\frac{\pi}{7} + i \sin (2k + 1)\frac{\pi}{7}, k = 0, 1, \ldots\ldots, 6$$

$$= \cos\left(\frac{p\pi}{7}\right) + i \sin\left(\frac{p\pi}{7}\right), p = 1, 3, 5, 7, 9, 11, 13$$

For p = 7, we get x = – 1, which is a root of x + 1 = 0.

Hence the six values of x are

$$\cos\left(\frac{p\pi}{7}\right) + i \sin\left(\frac{p\pi}{7}\right), p = 1, 3, 5, 9, 11, 13.$$

Example 9:

Solve $x^6 + x^5 + x^4 + x^3 + x^2 + x + 1 = 0.$

Solution :

Multiplying the given equation by (x – 1), we get

$$(x - 1)(x^6 + x^5 + x^4 + x^3 + x^2 + x + 1) = 0 \Rightarrow x^7 - 1 = 0.$$

Hence, by the required roots are

$$\cos\left(\frac{k\pi}{7}\right) \pm i \sin\left(\frac{k\pi}{7}\right), k = 2, 4, 6.$$

(Neglect the value x = 1 which corresponds to the factor x – 1).

Example 10:

Solve $x^9 + x^5 + x^4 + 1 = 0.$

Solution :

The given equation is $(x^5 + 1)(x^4 + 1) = 0.$

By the roots of $x^5 + 1 = 0$ are

$$\cos\left\{(2k+1)\frac{\pi}{5}\right\} + i \sin\left\{(2k+1)\frac{\pi}{5}\right\}, k = 0, 1, 2, 3, 4. \quad \ldots(A)$$

By the roots of $x^4 + 1 = 0$ are

$$\frac{1}{\sqrt{2}}\ (\pm 1 \pm i) \qquad ...(B).$$

Hence the values given in (A) and (B) are the required 9 roots of the given equation.

Example 11:

Solve $z^4 - z^3 + z^2 - z + 1 = 0.$

Solution :

Multiplying the given equation by $(z + 1)$ and simplying, we obtain, $2^5 + 1 = 0$, whose roots are

$z = \cos (p\pi/5) + \sin (p\pi/5)$, $p = 1, 3, 5, 7, 9,$

For $p = 5$, the root $z = -1$

corresponds to $z + 1 = 0$.

Hence the required roots are

$$\cos (p\pi/5) + i \sin (p\pi/5);\ p = 1, 3, 7, 9.$$

Example 12:

Solve $z^5 + z^4 + z^3 + z^2 + z + 1 = 0.$

Solution :

$z^5 + z^4 + z^3 + z^2 + z + 1 = (z^4 - 1)/(z - 1).$

By the roots of $z^6 - 1 = 0$ are

$$\cos (k\pi/3) + i \sin (k\pi/3);\ k = 0, 1, 2, 3, 4, 5.$$

Rejecting the root $z = 1$ (which corresponds to $k = 0$), the required roots are

$$\cos (k\pi/3) + i \sin (k\pi/3);\ k = 1, 2, 3, 4, 5.$$

$$\Rightarrow\ -1, \pm \frac{1}{2} \pm i\frac{\sqrt{3}}{2}.$$

Example 13:

Solve the equation $z^8 - z^5 + z^3 - 1 = 0.$

Solution :

$z^8 - z^5 + z^3 - 1 = (z^5 + 1)(z^3 - 1).$

Solving $z^3 - 1 = 0$ and $z^5 + 1 = 0$, the required roots are

$$-1, -\frac{1}{2}\left(1 \pm i\sqrt{3}\right);\ \cos (k\pi/5) + i \sin (k\pi/5),\ k = 1, 3, 5, 7, 9.$$

Example 14:

Solve the equation $z^{10} - 2^5 + 1 = 0$.

Solution :

Solving $z^{10} - z^5 + 1 = 0$ as a quadratic in z^5, we get

$$z^5 = \frac{1 \pm \sqrt{1-4}}{2} = \frac{1}{2} \pm i\frac{\sqrt{3}}{2} = \cos\frac{\pi}{3} \pm i\sin\frac{\pi}{3}$$

$$\Rightarrow \quad z = [\cos(2k\pi + \pi/3) \pm i\sin(2k\pi + \pi/3)]^{1/5}.$$

Hence $z = \cos(6k+1)\frac{\pi}{15} \pm i\sin(6k+1)\frac{\pi}{15}$, $k = 0, 1, 2, 3, 4$ are the required roots.

Example 15:

Solve the equation $x^6 + x^3 + 1 = 0$.

Solution :

Solving $x^6 + x^3 + 1 = 0$ as a quadratic in x^3, we get

$$x^3 = \frac{-1 \pm \sqrt{1-4}}{2} = -\frac{1}{2} \pm i\frac{\sqrt{3}}{\sqrt{2}} = r(\cos\theta \pm i\sin\theta).$$

$$\therefore \quad r\cos\theta = -\frac{1}{2} \text{ and } r\sin\theta = \sqrt{3/2}$$

$$\Rightarrow \quad r^2 = 1 \text{ and } \tan\theta = -\sqrt{3}.$$

$$\Rightarrow \quad r = 1 \text{ and } \theta = \frac{2\pi}{3} \qquad \Rightarrow x^3 = \cos\frac{2\pi}{3} \pm i\sin\frac{2\pi}{3}.$$

$$\therefore \quad x = \left[\cos\left(2kp + \frac{2\pi}{3}\right) \pm i\sin\left(2k + \frac{2\pi}{3}\right)\right]^{1/2}$$

$$= \cos(6k+2)\frac{\pi}{9} \pm i\sin(6k+2)\frac{\pi}{9}, \; k = 0, 1, 2$$

Hence $\cos\left(\frac{p\pi}{9}\right) \pm i\sin\left(\frac{p\pi}{9}\right)$, $p - 2, 8, 14$

are the required roots.

Example 16:

Solve the equation $z^4 + 4z^2 + 16 = 0$.

Solution :

Solving $z^4 + 4z^2 + 16 = 0$ as a quadratic in z^2,

$$z^2 = \frac{-4 \pm \sqrt{16-64}}{2} = -2 \pm 2\sqrt{3}i$$

Let $z^2 = -2 \pm 2\sqrt{3}i = r(\cos\theta \pm i\sin\theta)$. ...(1)

$\therefore$ $r\cos\theta = -2$ and $r\sin\theta = 2\sqrt{3}$.

$\Rightarrow$ $r^2 = 16$ and $\tan\theta = -\sqrt{3} \Rightarrow r = 4, \theta = 2\pi/3$. ...(2)

$$z = 2\left(\cos\frac{2\pi}{3} \pm i\sin\frac{2\pi}{3}\right)^{1/2}, \text{ by (1) and (2)}$$

$$= \left[\cos\left(2k\pi + \frac{2\pi}{3}\right) \cdot \frac{1}{2} \pm i\sin\left(2k\pi + \frac{2\pi}{3} \cdot \frac{1}{2}\right)\right]; k = 0, 1.$$

$$= 2\left(\cos\frac{\pi}{3} \pm i\sin\frac{\pi}{3}\right), 2\left(\cos\frac{4\pi}{3} \pm i\sin\frac{4\pi}{3}\right),$$

$$= \left(\frac{1}{2} \pm i\frac{\sqrt{3}}{2}\right), -2\left(\frac{1}{2} \pm i\frac{\sqrt{3}}{2}\right)$$

Hence $\pm 1 \pm i\sqrt{3}$ are the required roots.

Example 17:

Find the seven seventh roots of unity and prove that for any positive integer n the sum of their nth powers vanishes, unless n is a multiple of 7 in which case the sum is 7.

Solution :

The seven seventh roots of unity *i.e.,*

$$x = (1)^{1/7} = (\cos 2k\pi + i\sin 2k\pi)^{1/7}$$

are $\alpha^k \equiv \cos(2k\pi/7) + i\sin(2k\pi/7)$, $k = 0, 1, 2, \ldots, 6$.

We have $\alpha = \cos(2\pi/7) + i\sin(2\pi/7)$.

The sum of the nth powers of these roots is

$$S = 1^n + \alpha^n + (\alpha^2)^n + (\alpha^3)^n + (\alpha^4)^n + (\alpha^5)^n + (\alpha^6)^n$$

$\Rightarrow$ $S = 1 + \alpha^n + \alpha^{2n} + \alpha^{3n} + \alpha^{4n} + \alpha^{5n} + \alpha^{6n}$. ...(1)

Case I: Let n be a mutliple of 7 so that $n = 7m$, m being a positive integer.

$\therefore$ $\alpha^{kn} = (\alpha^k)^{7m} = \{\cos(2k\pi/7) + i\sin(2k\pi/7)\}^{7m}$

$= \cos 2km\pi + i\sin 2km\pi$

$= (\cos 2m\pi + i\sin 2m\pi)^k = 1^k$.

Thus $\alpha^{kn} = 1$ for $k = 0, 1, 2, \ldots\ldots, 6$.

Using in (1), S = 7.

Case II: Let n be not a multiple of 7 so that $\alpha^n \neq 1$.

From (1), $S = \dfrac{1-(\alpha^n)^7}{1-\alpha^n} = \dfrac{1-(\alpha^7)^n}{1-\alpha^n} = \dfrac{1-1}{1-a^n}$ $\quad (\because \alpha^7 = 1)$

Hence $\quad S = 0$, as $1 - a^n \neq 0$.

Example 18:

Show that the roots of the equation $(z-1)^5 + z^5 = 0$ *are given by*

$$z = \frac{1}{2}\left[1 + i\cos\left(\frac{p\pi}{10}\right)\right],\ p = 1, 3, 5, 7, 9.$$

Solution :

The given equation may be written as

$$\left(\frac{z-1}{z}\right)^5 + 1 = 0$$

$\Rightarrow w^5 + 1 = 0$, where

$$w = \frac{z-1}{z}$$

$$\Rightarrow\ z = \frac{1}{1-w}. \qquad ...(1)$$

Now $w^5 + 1 = 0$

$\Rightarrow w^5 = -1 = \cos\pi + i\sin\pi.$

$\Rightarrow\ w = (\cos\pi + i\sin\pi)^{1/5}.$

$$\Rightarrow\ w = \cos\left\{\frac{2k\pi + \pi}{5}\right\} + i\sin\left\{\frac{2kp + \pi}{5}\right\},\ k = 0, 1, 2, 3, 4.$$

We write $w = \cos\theta + i\sin\theta$, $\theta = (2k+1)\dfrac{\pi}{5}$.

Putting this value of of w in (1), we obtain

$$z = \frac{1}{1-\cos\theta - i\sin\theta}$$

$$= \frac{1}{1-\left(1-2\sin^2\theta/2\right) - 2i\sin\theta/2\cos\theta/2}$$

$$= \frac{1}{-2i\sin\theta/2\left\{\cos^2\theta/2 + i\sin\theta/2\right\}}$$

$$= \frac{(\cos\theta/2 - i\sin\theta/2)}{-2i\sin\theta/2} = \frac{1}{2} - \frac{1}{2i}\cot\frac{\theta}{2}$$

$$= \frac{1}{2}\,(1 + i\cot q/2) \qquad (\therefore\ -1/i = i^2/i = i)$$

$$= \frac{1}{2}\left[1 + i\cot\left\{\frac{(2k+1)\pi}{10}\right\}\right],\ k = 0, 1, 2, 3, 4.$$

$$= \frac{1}{2}\left[1 + i\cot\left(\frac{p\pi}{10}\right)\right],\ p = 1, 3, 5, 7, 9.$$

Example 19:

Show that the roots of the equation

$$(1 - z)^7 + (1 + z)^7 = 0 \text{ are } z = \pm i\cot\left(\frac{p\pi}{7}\right),\ p - 1, 2, 3.$$

Solution :

We have

$$\left(\frac{1+z}{1-z}\right)^7 + 1 = 0 \Rightarrow w^7 + 1 = 0, \text{ where}$$

$$w = \frac{1+z}{1-z} \Rightarrow z = \frac{w-1}{w-1}. \qquad ...(1)$$

Now $w^7 + 1 = 0 \Rightarrow w^7 = -1 = \cos\pi + i\sin\pi$

$\Rightarrow\ w = (\cos\pi + i\sin\pi)^{1/7}$.

$\therefore\ w = \cos\left\{(2k+1)\frac{\pi}{7}\right\} + i\sin\left\{(2k+1)\frac{\pi}{7}\right\}$, k = 0, 1, 2. 3, 4, 5,

6.

$\Rightarrow\ w = \cos\theta + i\sin\theta$, where $\theta = (2k+1)\frac{\pi}{7}$.

Putting this value of w in (1), we obtain

$$z = \frac{\cos\theta + i\sin\theta - 1}{1 + \cos\theta + i\sin\theta}$$

$$= \frac{1 - 2\sin^2\theta/2 + 2i\sin\theta/2\cos\theta/2 - 1}{1 + 2\cos^2\theta/2 - 1 + 2i\sin\theta/2\cos\theta/2}$$

$$= \frac{2i\sin\theta/2\{\cos\theta/2 + i\sin\theta/2\}}{2\cos\theta/2\{\cos\theta/2 + i\sin\theta/2\}}$$

$$= i \tan \frac{\theta}{2} = i \sin \left\{(2k+1)\frac{\pi}{14}\right\}, \; k = 0, 1, 2, 3, 4, 5, 6.$$

For k = 3, we get an infinite value of z which is rejected.

Thus $z = i \tan\left(\frac{p\pi}{14}\right)$, where p = 1, 3, 5, 9, 11, 13. ...(2)

We observe that

$$\tan\left(\frac{\pi}{14}\right) = \tan\left(\frac{\pi}{2} - \frac{3\pi}{7}\right) = \cot\frac{2\pi}{7};$$

$$\tan\left(\frac{3\pi}{14}\right) = \tan\left(\frac{\pi}{2} - \frac{2\pi}{7}\right) = \cot\frac{2\pi}{7};$$

$$\tan\left(\frac{5\pi}{14}\right) = \tan\left(\frac{\pi}{2} - \frac{\pi}{7}\right) = \cot\frac{\pi}{7}$$

$$\tan\left(\frac{9\pi}{14}\right) = \tan\left(\frac{\pi}{2} + \frac{\pi}{7}\right) = -\cot\frac{\pi}{7};$$

$$\tan\left(\frac{11\pi}{14}\right) = -\cot\frac{2\pi}{7}, \; \tan\left(\frac{13\pi}{4}\right) = -\cot\frac{3\pi}{7}.$$

Hence, by (2), the required roots are

$$z = \pm i \cot\left(\frac{p\pi}{7}\right), \; p = 1, 2, 3.$$

Example 20:

Solve the equation

$$(z+1)^7 = z^7 (\cos 7\theta + i \sin 7\theta).$$

Solution :

The given equation is $w^7 = \cos 7\theta + i \sin 7\theta$, ...(1)

$$w = \frac{z+1}{z} \Rightarrow z = \frac{1}{w-1}. \quad \text{From (1),}$$

$$w = \cos\left(\frac{2k\pi + 7\theta}{7}\right) + i \sin\left(\frac{2k\pi + 7\theta}{7}\right), \; k = 0, 1, ..., 6$$

$$\Rightarrow \; w = \cos\alpha + i \sin\alpha, \; \alpha = \theta + (2/7)\, k\pi.$$

$$\therefore \quad z = \frac{1}{\cos\alpha + i \sin\alpha - 1}$$

$$= \frac{1}{\{1-2\sin^2(\alpha/2)\} + 2i\sin(\alpha/2)\cos(\alpha/2) - 1}$$

$$= \frac{1}{2i\sin(\alpha/2)\{\cos(\alpha/2) + i\sin(\alpha/2)\}}$$

$$= \frac{\cos(\alpha/2) - i\sin(\alpha/2)}{2i\sin(\alpha/2)} = -\frac{1}{2} = \frac{1}{2i}\cot\left(\frac{\alpha}{2}\right)$$

$$= -\frac{1}{2} + \frac{1}{2i^2}\cot\left(\frac{\alpha}{2}\right) \quad (i^2 = -1)$$

$$= -\frac{1}{2}\left\{1 + i\cot\left(\frac{\alpha}{2}\right)\right\}$$

$$= -\frac{1}{2}\left\{1 + i\cot\left(\frac{\alpha}{2} + k\frac{\pi}{7}\right)\right\}, \ k = 0, 1 \ldots, 6.$$

Example 21:

Solve the equation $(2 + z)^6 + (2 - z)^6 = 0$.

Solution :

The given equation can be written as

$$w^6 + 1 = 0, \text{ where } w = \frac{2+z}{2-z} \qquad \ldots(1)$$

$$\Rightarrow \ w^6 = -1 = \cos\pi + i\sin\pi.$$

$$\therefore \quad w = \left\{\cos\left(\frac{2k\pi + \pi}{6}\right) + i\sin\left(\frac{2k\pi + \pi}{6}\right)\right\}, \ k = 0, 1, \ldots, 5$$

$$\Rightarrow \ w = \cos\theta + i\sin\theta, \text{ where } \theta = (2k+1)\frac{\pi}{6}.$$

From (1), $(2 - z)w = 2 + z \Rightarrow z = \dfrac{2(w-1)}{w+1}$

$$\Rightarrow \quad z = \frac{2(\cos\theta + i\sin\theta - 1)}{(\cos\theta + i\sin\theta + 1)}$$

$$= \frac{\cos(\alpha/2) - i\sin(\alpha/2)}{2i\sin(\alpha/2)} = -\frac{1}{2} = \frac{1}{2i}\cot\left(\frac{\alpha}{2}\right)$$

$$= -\frac{1}{2} + \frac{1}{2i^2}\cot\left(\frac{\alpha}{2}\right) \quad (i^2 = -1)$$

$$= -\frac{1}{2}\left\{1 + i\cot\left(\frac{\alpha}{2}\right)\right\}$$

$$= -\frac{1}{2}\left\{1 + i\cot\left(\frac{\alpha}{2} + k\frac{\pi}{7}\right)\right\},\ k = 0, 1....., 6.$$

Example 22:

Solve the equation $(2 + z)^6 + (2 - z)^6 = 0.$

Solution :

The given equation can be written as

$$w^6 + 1 = 0, \text{ where } w = \frac{2+z}{2-z} \qquad ...(1)$$

$$\Rightarrow\ w^6 = -1 = \cos\pi + i\sin\pi.$$

$$\therefore\quad w = \left\{\cos\left(\frac{2k\pi + \pi}{6}\right) + i\sin\left(\frac{2k\pi + \pi}{6}\right)\right\},\ k = 0, 1, ..., 5$$

$$\Rightarrow\ w = \cos\theta + i\sin\theta, \text{ where } \theta = (2k+1)\frac{\pi}{6}.$$

From (1), $(2 - z)w = 2 + z \Rightarrow z = \dfrac{2(w-1)}{w+1}$

$$\Rightarrow\quad z = \frac{2(\cos\theta + i\sin\theta - 1)}{(\cos\theta + i\sin\theta + 1)}$$

$$= \frac{2\left\{(1 - 2\sin^2\theta/2) + 2i\sin\theta/2\cos\theta/2 - 1\right\}}{\left\{(2\cos^2\theta/2 - 1) + 2i\sin\theta/2\cos\theta/2 + 1\right\}}$$

$$= \frac{2i\sin\theta/2\left\{\cos\theta/2 + i\sin\theta/2\right\}}{\cos\theta/2\left\{\cos\theta/2 + i\sin\theta/2\right\}} = 2i\tan\frac{\theta}{2}$$

Hence $z = 2i\tan\left\{(2k+1)\dfrac{\pi}{12}\right\},\ k = 0, 1, 2, ...5.$

Example 23:

Solve $(1 + iz)^{10} + (1 - iz)^{10} = 0.$

Solution :

The given equation can be expressed as

$$w^{10} + 1 = 0 \Rightarrow w = (-1)^{1/10} \Rightarrow w = (\cos \pi + i \sin \pi)^{1/10},$$

where $$w = \frac{1 + iz}{1 - iz} \Rightarrow z = \frac{w-1}{i\,(w+1)}$$

$$= \frac{i\,(w-1)}{i^2\,(w+1)} = \frac{i\,(1-w)}{1+w}.$$

We have

$$w = \cos\,(2k+1)\,\frac{\pi}{10} + i \sin\,(2k+1)\,\frac{\pi}{10},\ k = 0, 1, \ldots, 9.$$

$$\Rightarrow\ w = \cos\theta + i \sin\theta,\ \theta = (2k+1)\pi/10$$

$$\therefore\quad z = \frac{i(1-w)}{1+w} = \frac{i\,(1-\cos\theta - i\sin\theta)}{1+\cos\theta + i\sin\theta}$$

$$= \frac{i\,\{1-(1-2\sin^2\theta/2) - 2i\sin\theta/2\cos\theta/2\}}{1+2\cos^2\theta/2 - 1 + 2i\sin\theta/2\cos\theta/2}$$

$$= \frac{-2i^2\sin\theta/2\,(\cos\theta/2 + i\sin\theta/2)}{2\cos\theta/2\,(\cos\theta/2 + i\sin\theta/2)} = \tan\frac{\theta}{2}$$

$$= \tan\,(2k+1)\,\frac{\pi}{10};\ k = 0, 1, \ldots 9.$$

Hence $z = \pm \tan\,(p\pi/20)$, $k = 1, 3, 5, 7, 9$.

[$\because$ for $k = 5$, $z = \tan\,(11\pi/20) = \tan\,\{\pi - (\pi/20)\} = -\tan\,(9\pi/20)$ etc.]

Example 24:

Show that the roots of $(2 - iz)^8 - z^8 = 0$ are $z = -i$, and $-i \pm \tan(\pi/8)$, for $\pi = 1, 2, 3$.

Solution :

The given equation is expressible as $w^8 - 1 = 0$, where $w = \dfrac{2-iz}{z}$

$$\Rightarrow\ z = \frac{2}{w+i}.$$

Now $w^8 = 1 = \cos 0 + i \sin 0 \Rightarrow w = (\cos 2k\pi + i \sin 2k\pi)^{1/8}$

$$\therefore\quad w = \cos\,(k\pi/4) + i \sin\,(k\pi/4),\ k = 0, 1, \ldots, 7$$

$$\Rightarrow\ w = \cos\theta + i\sin\theta,\ \theta = k\pi/4$$

We have $$z = \frac{2}{w+i} = \frac{2}{\cos\theta + i\,(1+\sin\theta)}$$

$$= \frac{2\left\{\cos\theta - i\left(1+\sin\theta\right)\right\}}{\left\{\cos\theta + i\left(1+\sin\theta\right)\right\}\left\{\cos\theta - i\left(1+\sin\theta\right)\right\}}$$

$$= \frac{2\left\{\cos\theta - i\left(1+\sin\theta\right)\right\}}{\cos^2\theta + \left(1+\sin\theta\right)^2 = 2\left(1+\sin\theta\right)}$$

$$= -i + \frac{2\sin\left(\pi/4 - \theta/2\right)\cos\left(\pi/4 - \theta/2\right)}{1 + 2\cos^2\left(\pi/4 - \theta/2\right) - 1}$$

$$= -i + \tan\left(\frac{\pi}{4} - \frac{\theta}{2}\right) = -i + \tan\left(\frac{\pi}{4} - \frac{k\pi}{8}\right)$$

$-i$ (for $k = 2$); $-i + \tan(2\pi/8)$ (for $k = 0$),

$-i + \tan(\pi/8)$ (for $k = 1$),

$=$ $-i - \tan(\pi/8)$ for ($k = 3$), $-i - \tan(2\pi/8)$ (for k $= 4$),

$-i - \tan(3\pi/8)$ (for $k = 5$), $-i - \infty$ (for $k = 6$),

$-i + \tan(-5\pi/8)$ (for $k = 7$)

$= i - \tan(\pi - 3\pi/8) = -i + \tan(3\pi/8)$.

Hence the roots are $-i$, $-i \pm \tan(p\pi/8)$, $p = 1, 2, 3$, after rejecting the value for $k = 6$.

Example 25:

Prove that

(i) $\cos 5\theta = \cos^5\theta - 10\cos^3\theta\sin^2\theta + 5\cos\theta\sin^4\theta$.

(ii) $\sin 5\theta = 5\cos^4\theta\sin\theta - 10\cos^2\sin^3\theta + \sin^5\theta$.

Solution :

We have $\cos 5\theta + i\sin 5\theta = (\cos\theta + i\sin\theta)^5$

$= \cos^5\theta + {}^5C_1\cos^4\theta\,(i\sin\theta) + {}^5C_2\cos^3\theta\,(i\sin\theta)^2$

$+ {}^5C_3\cos^2\theta\,(i\sin\theta)^3 + {}^5C_4\cos\theta\,(i\sin\theta)^4 + (i\sin\theta)^5$

$$= \cos^5\theta + 5i\cos^4\theta\sin\theta - \frac{5.4}{2.1}\,3\cos^2\theta\sin^2\theta$$

$$- \frac{5.4.3}{3.2.1}\,i\cos^2\theta\sin^3\theta + \frac{5.4.3.2}{4.3.2.1}\cos\theta\sin^4\theta + i\sin^5\theta$$

$(\because i^2 = -1,\ i^3 = -i,\ i^4 = 1,\ i^5 = i \text{ etc.})$

Thus $\cos 5\theta + i\sin 5\theta = (\cos^5\theta - 10\cos^3\theta\sin^2\theta + 5\cos\theta\sin^4\theta)$

$$+ i(5\cos^4 \sin\theta - 10\cos^2\theta\sin^3\theta + \sin^5\theta).$$

Equating real and imaginary parts on both the sides, we get

$$\cos 5\theta = \cos^5\theta - 10\cos^3\theta\sin^2\theta + 5\cos\theta\sin^4\theta,$$

$$\sin 5\theta = 5\cos^4\theta\sin\theta - 10\cos^2\theta\sin^3\theta + \sin^5\theta.$$

Example 26:

Write down the expansions of cos 6θ, sin 6θ, tan 6θ.

Solution :

$$\cos 6\theta + i\sin 6\theta = (\cos\theta + i\sin\theta)^6$$

$$= \cos^6\theta + {}^6C_1\cos^5\theta\, i\sin\theta + {}^6C_2\cos^4\theta\, i^2\sin^2\theta + {}^6C_3\cos^3\theta\, i^2\sin^3\theta$$
$$+ {}^8C_4\cos^2\theta\, i^4\sin^4\theta + {}^6C_5\cos\theta\, i^5\sin^5\theta + {}^6C_6\, i^6\sin^6\theta.$$

Equating real and imaginary parts, we get

$$\cos 6\theta = \cos^6\theta - {}^6C_2\cos^4\theta\sin^2\theta + {}^6C_4\cos^2\theta\sin^4\theta - {}^6C_6\sin^6\theta$$
$$= \cos^6\theta - 15\cos^4\theta\sin^2\theta + 14\cos^2\theta\sin^4\theta - \sin^6\theta,$$

and $\sin 6\theta = {}^6C_1\cos^5\theta\sin\theta - {}^6C_3\cos^3\theta\sin^3\theta + {}^6C_5\cos\theta\sin^5\theta$

$$= 6\cos^5\theta\sin\theta - 20\cos^3\theta\sin^3\theta + 6\cos\theta\sin^5\theta.$$

Now $\tan 6\theta = \dfrac{\sin 6\theta}{\cos 6\theta}$

$$= \frac{6\cos^5\theta\sin\theta - 20\cos^3\theta\sin^3\theta + 6\cos\theta\sin^5\theta}{\cos^6\theta - 15\cos^4\theta\sin^2\theta + 15\cos^2\theta\sin^4\theta - \sin^6\theta}$$

On dividing numerator and denominator by $\cos^6\theta$, we obtain

$$\tan 6\theta = \frac{6\tan\theta - 20\tan^3\theta + 6\tan^5\theta}{1 - 15\tan^2\theta + 15\tan^4\theta - \tan^6\theta}.$$

Example 27:

Expand cos 7θ and sin 7θ in terms of powers of sin θ and cos θ.

Solution :

By De Moivre's Theorem, we have

$$(\cos\theta + i\sin\theta)^7 = \cos 7\theta + i\sin 7\theta$$

$$\Rightarrow \cos 7\theta + i\sin 7\theta = \cos^7\theta + {}^7C_1\cos^6\theta(i\sin\theta) + {}^7C_2\cos^5\theta(i\sin\theta)^2$$
$$+ {}^7C_3\cos^4\theta(i\sin\theta)^3 + {}^7C_4\cos^3\theta(i\sin\theta)^4$$
$$+ {}^7C_5\cos^2\theta(i\sin\theta)^5 + {}^7C_6\cos\theta(i\sin\theta)^6$$
$$+ {}^7C_7(i\sin\theta)^7$$

$$= \cos^7\theta + 7i\cos^6\theta\sin\theta - 21\cos^5\theta\sin^2\theta - 35i\cos^4\theta\sin^3\theta$$

$+ 35 \cos^3\theta \sin^4\theta + 21i \cos^2\theta \sin^5\theta - 7 \cos\theta \sin^6\theta - i \sin^7\theta$

$= (\cos^7\theta - 21 \cos^5\theta \sin^2\theta + 35 \cos^3\theta \sin^4\theta - 7 \cos\theta \sin^6\theta)$

$+ i (7 \cos^6\theta \sin\theta - 35 \cos^4\theta \sin^3\theta + 21 \cos^2\theta \sin^5\theta - \sin^7\theta).$

Equating the real and imaginary parts on both sides, we get

$\cos 7\theta = \cos^7\theta - 21 \cos^5\theta \sin^2\theta + 35 \cos^3\theta \sin^4\theta - 7 \cos\theta \sin^6\theta,$

$\sin 7\theta = 7 \cos^6\theta \sin\theta - 35 \cos^4\theta \sin^3\theta + 21 \cos^2\theta \sin^5\theta - \sin^7\theta.$

EXPANSIONS OF $\text{COS}^N\theta$ AND $\text{SIN}^N\theta$ IN TERMS OF SINES AND COSINES OF MULTIPLES OF θ, N BEING A POSITIVE INTEGER

Let $z = \cos\theta + i \sin\theta$ so that $\frac{1}{z} = \cos\theta - i \sin\theta$.

Now $z + \frac{1}{z} = 2 \cos\theta$ and $z - \frac{1}{z} = 2i \sin\theta$.

By De Moivre's Theorem, we have

$z^n = \cos n\theta + i \sin n\theta$ and $\frac{1}{z^n} = \cos n\theta - i \sin n\theta$.

Thus $z^n + \frac{1}{z^n} = 2 \cos n\theta$ and $z^n - \frac{1}{z^n} = 2i \sin n\theta$. ...(2)

From (2), we have

$$(2 \cos\theta)^n = \left(z + \frac{1}{z}\right)^n.$$

Expanding the R.H.S. with the help of the Binomial Theorem and using (2), we get the required expansion of $\cos^n\theta$. Similarly

$$(2i \sin\theta)^n = \left(z - \frac{1}{z}\right)^n$$

gives the required expansion of $\sin^n\theta$.

Example 1:

Expand $\sin^7\theta \cos^2\theta$ in a series of sines of multiplies of θ.)

Solution :

Let $z = \cos\theta + i \sin\theta$.

$$\therefore \quad (2i \sin\theta)^7 (2 \cos\theta)^3 = \left(z - \frac{1}{z}\right)^7 \left(z + \frac{1}{z}\right)^3$$

$$= \left(z - \frac{1}{z}\right)^4 \left(z^2 - \frac{1}{z^2}\right)^3$$

$$= (z^4 - 4z^2 + 6 - 4z^{-2} + z^{-4})(z^6 - 3z^2 + 3z^{-2} - z^{-6})$$

$$= z^{10} - 4z^8 + 3z^6 + 8z^4 - 14z^2 + 14z^{-8} - 8z^{-4} - 3z^{-6} + 4z^{-8} - z^{-10}$$

$$= (z^{10} - z^{-10}) - 4(z^8 - z^{-8}) + 3(z^6 - z^{-6})$$

$$+ 8(z^4 - z^{-4}) - 14(z^2 - z^{-2})$$

$$= 2i\,[\sin 10\theta - 4\sin 8\theta + 3\sin 6\theta + 8\sin 4\theta - 14\sin 2\theta].$$

Hence $-2^9 \sin^7\theta \cos^2\theta = \sin 10\theta - 4\sin 8\theta + 3\sin 6\theta$

$$+ 8\sin 4\theta - 14\sin 2\theta.$$

Example 2:

Prove that

$$-128\sin^6\theta\cos^2\theta = \cos 8\theta - 4\cos 6\theta + 4\cos 4\theta + 4\cos 2\theta - 5.$$

Solution :

Let $z = \cos\theta + i\sin\theta$ so that $1/z = \cos\theta - i\sin\theta$.

Thus $z + 1/z = 2\cos\theta$ and $z - 1/z = 2i\sin\theta$.

$$\therefore \quad (2i\sin\theta)^6 (2\cos\theta)^2 = \left(z - \frac{1}{z}\right)^6 \left(z - \frac{1}{z}\right)^2$$

$$= \left(z - \frac{1}{z}\right)^4 \left(z^2 - \frac{1}{z^2}\right)^2$$

$$= \left(z^4 - 4z^8 . \frac{1}{z} + 6z^2 . \frac{1}{z^2} - 4z . \frac{1}{z^3} + \frac{1}{z^4}\right)\left(z^4 - 2 + \frac{1}{z^4}\right)$$

$$= \left(z^8 + \frac{1}{z^8}\right) - 4\left(z^6 + \frac{1}{z^6}\right) + 4\left(z^4 + \frac{1}{z^4}\right) + 4\left(z^2 + \frac{1}{z^2}\right) + 10$$

$$= 2\cos 8\theta - 4(2\cos 6\theta) + 4(2\cos 4\theta) + 4(2\cos 2\theta) - 10$$

$$\Rightarrow -2^8 \sin^6 \cos^2\theta = 2(\cos 8\theta - 4\cos 6\theta + 4\cos 4\theta + 4\cos 2\theta - 5)$$

Hence $-128\sin^6\theta\cos^2\theta = \cos 8\theta - 4\cos 6\theta + 4\cos 4\theta$

$$+ 4\cos 2\theta - 5.$$

Example 3:

Prove that

$$128\sin^2\theta\cos^6\theta = -1\cos 8\theta - 4\cos 6\theta - 4\cos 4\theta + 4\,ccs\,2\theta + 5.$$

Solution :

Let $z = \cos\theta + i\sin\theta$ so that $z^{-1} = \cos\theta - i\sin\theta$.

$\therefore \quad z + z^{-1} = 2\cos\theta,\ z - z^{-1} = 2i\sin\theta.$

We have

$(2i \sin \theta)^2 (2 \cos \theta)^6 = (z - z^{-1})^2 (z + z^{-1})^6$

$= (z - z^{-1})^2 (z + z^{-1})^2 (z + z^{-1})^4$

$= (z^2 - z^{-2})^2 (z^4 + 4c_1 z^3 z^{-1} + 4c_3 z^2 z^{-2} + 4c_3 z z^{-3} + z^{-4})$

$= (z^4 - 2 + z^{-4}) (z^4 + 4z^2 + 6 + 4z^{-2} + z^{-4})$

$= (z^8 + z^{-8}) + 4 (z^6 + z^{-6}) + 4 (z^4 + z^{-4}) - 4 (z^2 + z^{-2}) - 10$

$\therefore \quad - 2^8 \sin^2 \theta \cos^6 \theta = 2 \cos 8\theta + 4 (2 \cos 6\theta)$

$+ 4 (2 \cos 4\theta) - 4 (2 \cos 2\theta) - 10.$

Hence

$128 \sin^2 \theta \cos^6 \theta = - \cos 8\theta - 4 \cos 6\theta - 4 \cos 4\theta + 4 \cos 2\theta + 5.$

Example 4:

Prove that

$256 \sin^5 \theta \cos^4 \theta = \sin 9\theta - \sin 7\theta - 4 \sin 5\theta + 4 \sin 3\theta + 6 \sin \theta.$

Solution :

We have

$(2i \sin \theta)^5 (2 \cos \theta)^4 = (z - z^{-1})^5 (z + z^{-1})^4$

$= (z^2 - z^{-2}) (z - z^{-1})$

$= (z^8 - 4z^4 + 6 - 4z^{-4} + z^{-8}) (z - z^{-1})$

$= (z^9 - z^{-9}) - (z^7 - z^{-7}) - 4 (z^5 - z^{-5}) + 4 (z^3 - z^{-3}) + 6 (z - z^{-1})$

$\therefore\ i\, 2^9 \sin^5 \theta \cos^4 \theta = 2\, i \sin 9\theta - 2i \sin 7\theta - 4 (2i \sin 5\theta)$

$+ 4 (2i \sin 3\theta) + (2i \sin \theta).$

Hence

$256 \sin^5 \theta \cos^4 \theta = \sin 9\theta - \sin 7\theta - 4 \sin 5\theta + 4 \sin 3\theta + 6 \sin \theta.$

Example 5:

Prove that

(i) $16 \sin^5 \theta = \sin 5\theta - 5 \sin 3\theta + 10 \sin \theta.$

(ii) $16 \cos^5 \theta = \cos 5\theta + 5 \cos 3\theta + 10 \cos \theta.$

Solution :

Let $z = \cos \theta + i \sin \theta$ so that $1/z = \cos \theta - i \sin \theta.$

$\Rightarrow \quad z - \dfrac{1}{z} = 2i \sin \theta$

$\Rightarrow \quad (2i \sin \theta)^5 = \left(z - \dfrac{1}{z}\right)^5$

$$= z^5 - {}^5C_1\, z^4.\frac{1}{z} + {}^5C_2\, z^3.\frac{1}{z^2} - {}^5C_3 z^2.\frac{1}{z^3} + {}^5C_4 z.\frac{1}{z^4} - \frac{1}{z^5}$$

$$= z^5 - 5\, z^3 + \frac{5.4}{2.1} z - \frac{5.4.3}{3.2.1}.\frac{1}{z} + \frac{5.4.3.2}{4.3.2.1}.\frac{1}{z^3} - \frac{1}{z^5}$$

$$= \left(z^5 - \frac{1}{z^5}\right) - 5\left(z^3 - \frac{1}{z^3}\right) + 10\left(z - \frac{1}{z}\right).$$

$= (2i \sin 5\theta) - 5\ (2i \sin 3\theta) + 10\ (2i \sin \theta)$, by (2) above

$\Rightarrow\ 32i \sin^5 \theta = 2i\ (\sin 5\theta - 5 \sin 3\theta + 10 \sin \theta)$ $\quad (\because\ i^5 = i)$

Hence $16 \sin^5 \theta = \sin 5\theta - 5 \sin 3\theta + 10 \sin \theta$.

(ii) We have $z + 1/z = 2 \cos \theta$, and so

$$(2 \cos \theta)^5 = \left(z + \frac{1}{z}\right)^5 = z^5 + 5z^3 + 10z + 10\ \frac{1}{z} + 5\ \frac{1}{z^3} + \frac{1}{z^5},$$

(as shown in the first part)

$$= \left(z^5 + \frac{1}{z^5}\right) + 5\left(z^3 + \frac{1}{z^3}\right) + 10\left(z + \frac{1}{z}\right)$$

$\Rightarrow\ 2^5 \cos^5 \theta = (2 \cos 5\theta) + 5\ (2 \cos 3\theta) + 10\ (2 \cos \theta)$, by (2) above.

Hence $16 \cos^5 \theta = \cos 5\theta + 5 \cos 3\theta + 10 \cos \theta$.

Example 6:

Prove that

$$32 \cos^6 \theta = \cos 6\theta + 6 \cos 4\theta + 15 \cos 2\theta + 10.$$

Solution :

Let $x = \cos \theta + i \sin \theta$ so that $1/x = \cos \theta - i \sin \theta$.

Thus $2 \cos \theta = x + \dfrac{1}{x} \Rightarrow (2 \cos \theta)^6 = \left(x + \dfrac{1}{x}\right)^6$

$$= x^6 + {}^6C_1 x^5\ .\ \frac{1}{x} + {}^6C_2 x^4.\frac{1}{x^2} + {}^6C_3 x^3.\frac{1}{x^3}$$

$$+ {}^6C_4 x^2.\frac{1}{x^4} + {}^6C_5 x.\frac{1}{x^5} + \frac{1}{x^6}$$

$$= x^6 + 6x^4 + 15x^2 + + 20 + \frac{15}{x^2} + \frac{6}{x} + \frac{1}{x^6}$$

$$= \left(x^6 + \frac{1}{x^6}\right) + 6\left(x^4 + \frac{1}{x^4}\right) + 15\left(x^2 + \frac{1}{x^2}\right) + 20$$

$\therefore$ 64 $\cos^6 \theta = 2 \cos 6\theta + 6.2 \cos 4\theta + 15.2 \cos 2\theta + 20$

Hence 32 $\cos^6 \theta = \cos 6\theta + 6 \cos 4\theta + 15 \cos 2\theta + 10$.

Example 7:

Prove that

$$64 \cos^7 \theta = \cos 7\theta + 7 \cos 5\theta + 21 \cos 3\theta + 35 \cos \theta.$$

Solution :

Let $x = \cos \theta + i \sin \theta$, then $1/x = \cos \theta - i \sin \theta$.

$$\therefore \quad (2 \cos \theta)^7 = \left(x+\frac{1}{x}\right)^7 = x^7 + 7c_1.x^6 . \frac{1}{x} + 7\, c_2.x^5 . \frac{1}{x^2}$$

$$+ 7c_3x^4.\frac{1}{x^3} + 7c_4.x^3.\frac{1}{x^4} + 7c_5x^2. + 7c_6x.\frac{1}{x^6} + 7c_7.\frac{1}{x^7}$$

$$= x^7 + 7x^5 + 21x^3 + 35x + 35.\frac{1}{x} + 21.\frac{1}{x^3} + 7.\frac{1}{x^5} + \frac{1}{x^7}$$

$$= \left(x^7+\frac{1}{x^7}\right) + 7\left(x^5+\frac{1}{x^5}\right) + 21\left(x^3+\frac{1}{x^3}\right) + 35\left(x+\frac{1}{x}\right)$$

$$= 2\, [\cos 7\theta + 7 \cos 5\theta + 21 \cos 3\theta + 35 \cos \theta].$$

Hence 64 $\cos^7 \theta = \cos 7\theta + 7 \cos 5\theta + 21 \cos 3\theta + 35 \cos \theta$.

Example 8:

Prove that

$$\sin^8 \theta = \frac{1}{128} \{\cos 8\theta - 8 \cos 6\theta + 28 \cos 4\theta - 56 \cos 2\theta + 70\}.$$

Solution :

Let $z = \cos \theta + i \sin \theta$

$\Rightarrow z^{-1} = \cos \theta - i \sin \theta$

$\Rightarrow \quad z - z^{-1} = 2\, i \sin \theta$

$$\Rightarrow \sin^8 \theta = \frac{1}{(2i)^8} (z - z^{-1})^8$$

$$= \frac{1}{256} (z^8 - {}^8c_1\, z^6 + {}^8c_2\, z^4 - {}^8c_3\, z^2 + {}^8c_4 - {}^8c_5\, z^{-2} + {}^8c_6\, z^{-4} - {}^8c_7\, z^{-6} + z^{-8})$$

$$= \frac{1}{256} \{(z^8 + z^{-8}) - 8\, (z^6 + z^{-6}) + 28\, (z^4 + z^{-4}) - 56\, (z^2 + z^{-2}) + 70\}$$

$$= \frac{1}{128} \{\cos 8\theta - 8 \cos 6\theta + 28 \cos 4\theta - 56 \cos 2\theta + 70\}.$$

Example 9:

Show that

$$\sin^6\theta = -\frac{1}{32}\,[\cos 6\theta - 6\cos 4\theta + 15\cos 2\theta - 10].$$

Solution :

We have

$$(2i\sin\theta)^6 = \left(x - \frac{1}{x}\right)^6$$

$$= x^6 + 6c_1x^5\left(-\frac{1}{x}\right) + 6c_2x^4\left(-\frac{1}{x}\right)^2$$

$$+ 6c_3x^3\left(-\frac{1}{x}\right)^3 + 6c_4x^2\left(-\frac{1}{x}\right)^4 + 6c_5x^2\left(-\frac{1}{x}\right)^5 + 6c_6\left(-\frac{1}{x}\right)^6$$

$$= x^6 - 6x^4 + 15x^2 - 20 + 15\,\frac{1}{x^2} - 6\,\frac{1}{x^4} + \frac{1}{x^6}$$

$$= \left(x^6 + \frac{1}{x^6}\right) - 6\left(x^4 + \frac{1}{x^4}\right) + 15\left(x^2 + \frac{1}{x^2}\right) - 20$$

$$= [2\cos 6\theta] - 6\,[2\cos 4\theta] + 15\,[2\cos 2\theta] - 20$$

Hence $\sin^6\theta = -\frac{1}{32}\,[\cos 6\theta - 6\cos 4\theta + 15\cos 2\theta - 10]$

Example 10:

Prove that

$$32\sin^4\theta\cos^2\theta = \cos 6\theta - 2\cos 4\theta - \cos 2\theta + 2.$$

Solution :

Let $x = \cos\theta + i\sin\theta$

$\Rightarrow 1/x = \cos\theta - i\sin\theta.$

$\Rightarrow x + 1/x = 2\cos\theta$ and $x - 1/x = 2i\sin\theta,$

$$\Rightarrow (2i\sin\theta)^4(2\cos\theta)^2 = \left(x + \frac{1}{x}\right)^2\left(x - \frac{1}{x}\right)^4$$

$$= \left(x + \frac{1}{x}\right)^3\left(x - \frac{1}{x}\right)^2\left(x - \frac{1}{x}\right)^2$$

$$= \left(x^2 - \frac{1}{x^2}\right)^2\left(x - \frac{1}{x}\right)^2$$

$$= \left(x^4 - 2 + \frac{1}{x^4}\right)\left(x^2 - 2 + \frac{1}{x^2}\right)$$

$= x^6 - 2x^4 - x^2 + 4 - \frac{1}{x^2} - \frac{2}{x^4} + \frac{1}{x^6}$

$= \left(x^6 + \frac{1}{x^6}\right) - 2\left(x^4 + \frac{1}{x^4}\right) - \left(x^2 + \frac{1}{x^2}\right) + 4$

$= 2 \cos 6\theta - 2.2 \cos 4\theta - 2 \cos 2\theta + 4.$

Hence $32 \sin^4 \theta \cos^2 \theta = \cos 6\theta - 2 \cos 4\theta - \cos 2\theta + 2.$

Example 11:

Prove that

$128 \cos^3 \theta \sin^5 \theta = \sin 8\theta - 2 \sin 6\theta - 2 \sin 4\theta + 6 \sin 2\theta.$

Solution :

If $z = \cos \theta + i \sin \theta$, then $z^{-1} = \cos \theta - i \sin \theta.$

$\Rightarrow \; 2 \cos \theta = z + z^{-1}, 2i \sin \theta = z - z^{-1}.$

$\Rightarrow \; (2 \cos \theta)^3 (2i \sin \theta)^5 = (z + z^{-1})^3 (z - z^{-1})^5$

$= (z^2 - z^{-2})^3 (z - z^{-1})^2$

$= (z^6 - 3z^2 + 3z^{-2} - z^{-6}) (z^2 - 2 + z^{-2})$

$= (z^8 - z^{-8}) - 2 (z^6 - z^{-6}) - 2 (z^4 - z^{-4}) + 6 (z^2 - z^{-2})$

$2^8 i \cos^3 \theta \sin^5 \theta = 2i \sin 8\theta - 2 (2i \sin 6\theta) - 2 (2i \sin 4\theta) + 6 (2i \sin 2\theta)$

Hence $128 \cos^3 \theta \sin^5 \theta = \sin 8\theta - 2 \sin 6\theta - 2 \sin 4\theta + 6 \sin 2\theta.$

Example 12:

Prove that

$64 \sin^3 \theta \cos^4 \theta = - \sin 7\theta - \sin 5\theta + 3 \sin 3\theta + 3 \sin \theta.$

Solution :

$(2 \cos \theta)^4 (2i \sin \theta)^3 = (z + z^{-1})^4 (z - z^{-1})^3$

$= (z^4 + 4z^2 + 6 + 4z^{-2} + z^{-4}) (z^3 - 3z + 3z^{-1} - z^{-3})$

$= (z^7 - z^{-7}) + (z^5 - z^{-5}) - 3 (z^3 - z^{-3}) -3 (z - z^{-1})$

$\Rightarrow \; - 2i (64 \sin^3 \theta \cos^4 \theta)$

$= 2i \sin 7\theta + 2i \sin 5\theta - 3 (2i \sin 3\theta) - 3 (2i \sin \theta)$

Hence $64 \sin^2 \theta \cos^4 \theta = - \sin 7\theta - \sin 5\theta + 3 \sin 3\theta + 3 \sin \theta.$

4

TRIGONOMETRICAL EXPANSION

INTRODUCTION TO GEOGORY'S SERIES

To prove that, if θ lies within the closed interval $[-\pi/4, \pi/4]$, i.e., $-\pi/4 \le \theta \le \pi/4$, then

$$\theta = \tan\theta - \frac{1}{3}\tan^3\theta + \frac{1}{5}\tan^5\theta - \frac{1}{7}\tan^7\theta + \ldots ad.inf.$$

Proof: We have

$$(1 + i\tan\theta) = \left(1 + i\frac{\sin\theta}{\cos\theta}\right) = \frac{1}{\cos\theta}(\cos\theta + i\sin\theta)$$

$$= \sec\theta.e^{i\theta}.$$

Taking logarithim of both sides, we have

$\log(1 + i\tan\theta) = \log\sec\theta + \log e^{i\theta}$, (considering only principal values)

or $\qquad \log(1 + i\tan\theta) = \log\sec\theta + i\theta.$

Now since θ lies between $-\pi/4$ and $\pi/4$, $\tan\theta$ lies between -1 and 1, *i.e.*, $\tan\theta$ is numerically not greater than unity.

Therefore, $\log\sec\theta + i\theta = \log(1 + i\tan\theta)$

$$= i\tan\theta - \frac{i^2\tan^2\theta}{2} + \frac{i^3\tan^3\theta}{3} - \frac{i^4\tan^4\theta}{4} + \ldots, \quad \ldots(1)$$

(expanding the R.H.S. by logarithmic series which is justified because $|i\tan\theta| = |\tan\theta| \le 1$)

$$= i\tan\theta + \frac{1}{2}\tan^2\theta - \frac{1}{3}i\tan^3\theta - \frac{1}{4}\tan^4\theta + \frac{1}{5}i\tan^5\theta + \ldots$$

Equating the imaginary parts on both sides, we have

$$\theta = \tan\theta - \frac{1}{3}\tan^3\theta + \frac{1}{5}\tan^5\theta - \ldots \quad \ldots(2)$$

The general term on the R.H.S. is $\frac{(-1)^{n-1}}{2n-1}\tan^{2n-1}\theta$.

This expansion (2) is known as Gregory's series after the name of Jamess Gregory (1638-1675).

Another Form of Gregory's Series

In the series (2) if we put tan θ = x o that θ = $\tan^{-1}$ x, then we have another form of the Gregory's series as

$$\tan^{-1}x = x - \frac{x^3}{3} + \frac{x^5}{5} - \frac{x^7}{7} + \text{...ad.inf. where } -1 \le x \le 1 \text{ } i.e., |x| \le 1.$$

Cor. Equating real parts from both sides of (1), we have

$$\log\sec\theta = \frac{1}{2}\tan^2\theta - \frac{1}{4}\tan^4\theta + \frac{1}{6}\tan^6\theta - ...$$

GENERAL THEOREM OF GREGORY'S SERIES

If q lies between $n\pi - \frac{1}{4}\pi$ *and* $n\pi + \frac{1}{4}\pi$*, both limits being inclusive, then*

$$\theta - n\pi = \tan\theta - \frac{1}{3}\tan^3\theta + \frac{1}{5}\tan^5\theta - \text{... ad.inf.}$$

Proof: Let θ – nπ = ϕ *i.e.*, θ = nπ + ϕ.

Then f lies between –π/4 and π/4.

Now 1 + i tan θ = 1 + i tan (nπ + ϕ) = 1 + i tan ϕ = sec ϕ (cos ϕ + i sin ϕ) = sec ϕ.$e^{i\phi}$.

Taking logarithm of both sides, we get

log (1 + i tan θ) = log sec ϕ + log $e^{i\phi}$

or log sec ϕ + iϕ = log (1 + i tan θ),

the expansion of which is valid because θ lies between $n\pi - \frac{1}{4}\pi$ and $n\pi + \frac{1}{4}\pi$ implies that tan θ is not numerically greater than 1.

$$\therefore \log\sec\phi + i\phi = i\tan\theta - \frac{1}{2}i^2\tan^2\theta + \frac{1}{3}i^3\tan^3\theta - \frac{1}{4}i^4\tan^4\theta + ...\infty$$

$$\left[\because \log(1+z) = z - \frac{1}{2}z^2 + \frac{1}{3}z^3 - ...\infty, \text{ if } |z| \le 1\right]$$

$$= i\tan\theta - \frac{1}{2}\tan^2\theta - \frac{1}{3}i\tan^3\theta - \frac{1}{4}\tan^4\theta + ...\infty.$$

Equating imaginary parts from both sides, we have

$$\phi = \tan\theta - \frac{1}{3}\tan^3\theta + \frac{1}{5}\tan^5\theta - \ldots \text{ad.inf.}$$

$$\therefore\ \theta - n\pi = \tan\theta - \frac{1}{3}\tan^3\theta + \frac{1}{5}\tan^5 - \ldots \text{ad.inf}$$

EXPANSION OF SOME TRIGONOMETRICAL FUNCTIONS

In this chapter, we shall expand some trigonometrical functions like $\cos^n\Theta$, $\sin^n\Theta$, $\cos^n\Theta\sin^m\Theta$, where n, m are positive integers. The following result will be used in this chapter.

Let $x = \cos\Theta + \sin\Theta$

Then $\frac{1}{x} = \frac{1}{\cos\Theta + i\sin\Theta}$

$= (\cos\Theta + i\sin\Theta)^{-1} = \cos\Theta - i\sin\theta$ [by De Moivre's Theorems]

$$\therefore\quad x + \frac{1}{x} = 2\cos\Theta$$

and $x - \frac{1}{x} = 2\sin\Theta$

By De Moivre'ss Theorem, we have

$$x^n = \cos n\theta + i\sin n\Theta$$

and $\frac{1}{x^n} = \cos n\theta - i\sin m\Theta$

$$\therefore\quad x^n + \frac{1}{x^n} = 2\cos m\Theta$$

and $x^n - \frac{1}{x^n} = 2i\sin n\Theta$.

EXPANSION OF COS^N Θ IN A SERIES OF COSINES OF MULTIPLE OF Θ, N BEING A POSITIVE INTEGER

We have $2\cos\theta = x + 1/x$

$\Rightarrow\quad (2\cos\theta)^n = (x + 1/x)^n$

$$\Rightarrow\quad 2^n\cos^n\theta = x^n + {}^nc_1\, x^{n-1}\frac{1}{x} + {}^nc_2\, x^{n-2}\frac{1}{x^2} + {}^nc_3\, x^{n-3}\frac{1}{x^3} + \ldots$$

$$\ldots + {}^nc_{n-2}.x^2\frac{1}{x^{n-2}} + {}^nc_{n-1}\, x.\frac{1}{x^{n-1}} + {}^nc_n\frac{1}{x^n}$$

But ${}^nc_r = {}^nc_{n-r}$, ${}^nc_1 = {}^nc_{n-1}$ etc. and ${}^nc_n = 1$

$\therefore\ 2^n\cos^n\theta = x^n + {}^nc_1\, x^{n-2} + {}^nc_2\, x^{n-4} + n^{n-4} + {}^nc_3\, x^{n-6} + \ldots.$

$$\ldots + {}^nc_2 \frac{1}{x^{n-4}} + {}^nc_1 \frac{1}{x^{n-2}} + \frac{1}{x^n} \qquad \ldots(i)$$

$$= \left(x^n + \frac{1}{x^n}\right) + {}^nc_1 \left(x^{n-2} + \frac{1}{x^{n-2}}\right) + {}^nc_2 \left(x^{n-4} + \frac{1}{x^{n-4}}\right) + \ldots,$$

arranging the terms taking 1st and last, 2nd and last but one and soon.

$$\Rightarrow\ 2^n \cos^n \theta = (2 \cos n\theta) + {}^nc_1 [2 \cos (n-2)\theta]$$
$$+ {}^nc_2 [2 \cos (n-4)\theta] + \ldots, \qquad \ldots(ii)$$

substituting the values of $x^n + \dfrac{1}{x^n}$ etc.

$$\Rightarrow\ 2^{n-1} \cos^n \theta = \cos n\theta + {}^nc_1 \cos (n-2)\theta + {}^nc_2 \cos (n-4)\theta + \ldots \qquad \ldots(iii)$$

If n be odd, the number of terms in R.H.S. of (i), will be (n + 1) which is even, so that all the terms form pairs, and the two middle terms being $\dfrac{n+1}{2}$th and $\dfrac{n+3}{2}$th will form the last pair

$$\frac{{}^nc_{n-1}}{2}\left(x+\frac{1}{x}\right), \text{ since } \frac{{}^nc_{n+1}}{2} = \frac{{}^nc_{n-1}}{2},$$

Therefore the last term in (ii) $= \dfrac{{}^nc_{n-1}}{2}\ (2 \cos \theta)$

$$\Rightarrow \text{ the last term in (iii)} = \frac{{}^nc_{n-1}}{2} \cos \theta = \frac{n!\cos q}{\{(n-1)/2!\}\{(n+1)/2!\}}$$

Again if n be even, the number of terms in R.H.S. of (i) will be (n + 1) which is odd, so that all the terms form pair except the middle one in (i), which is independent of x and is equal to ${}^nc_{n/2}$ as it is the $\left(\dfrac{1}{2}n+1\right)$ th term in the expansion on R.H.S. of (i). Hence the last term in (iii)

$$= {}^nc_{n/2} = \frac{n}{(n/2)!\,(n/2)!} = \frac{n!}{\{(n/2)!\}^2}$$

Example 1:

Expand $\cos^6 \theta$.

Solution:

If $x = \cos \theta + i \sin \theta$, then $1/x = \cos \theta - i \sin \theta$;

$x^n = \cos n\theta + i \sin n\theta$ and $1/x^n = \cos n\theta - i \sin n\theta$.

Hence $x + (1/x) = 2 \cos \theta$ and $x^n + (1/x^n) = 2 \cos n\theta$.

Now $(2 \cos \theta)^6 = \left(x+\frac{1}{x}\right)^6$

$$\Rightarrow \quad 2^6 \cos^8 \theta = x^6 + 6x^4 + 15x^2 + 20 + 15.\frac{1}{x^2} + 6\frac{1}{x^4} + \frac{1}{x^6}$$

$$\Rightarrow \quad 2^6 \cos^6 \theta = \left(x^6+\frac{1}{x^6}\right) + 6\left(x^4+\frac{1}{x^4}\right) + 15\left(x^2+\frac{1}{x^2}\right) + 20$$

$$= (2 \cos 6\theta) + 6\cdot(2 \cos 4\theta) + 15\,(2 \cos 2\theta) + 20,$$

$$\therefore \quad x^n + \frac{1}{x^n} = 2 \cos n\theta$$

$2^5 \cos^6 \theta = \cos 6\theta + 6 \cos 4\theta + 15 \cos 2\theta + 10$ *(Kashmir 74)*

$$\Rightarrow \quad \cos^6 \theta = \frac{1}{22} \cos 6\theta + \frac{33}{16} \cos 4\theta + \frac{15}{22} \cos 2\theta + \frac{5}{16}$$ **Ans.**

Example 2:

Expand $\cos^9 \theta$.

Solution:

With the same o'ratio as in above, we have $(2 \cos \theta)^9 = (x + 1/x)^9$

$$= x^9 + 9c_1\, x^8\, (1/x) + 9c_2\, x^7\, (1/x)^2 + 9c_3\, x^6\, (1/x)^3$$
$$+ 9c_4 x^5\, (1/x)^4 + 9c_5 x^4\, (1/x)^5 + 9c_0 x^3\, (1/x)^6$$
$$+ 9c_7 x^2\, (1/x)^7 + 9c_8 x\, (1/x)^8 + 9c_9\, (1/x)^9$$

$$= x^9 + 9x^7 + 36x^5 + 84x^3 + 126x + \frac{126}{x} + \frac{84}{x^5} + \frac{36}{x^5} + \frac{9}{x^7} + \frac{1}{x^9}$$

$$= \left(x^9+\frac{1}{x^9}\right) + 9\left(x^7+\frac{1}{x^7}\right) + 36\left(x^5+\frac{1}{x^5}\right)$$
$$+ 84\left(x^3+\frac{1}{x^3}\right) + 126\left(x+\frac{1}{x}\right)$$

$$= (2 \cos 9\theta) + 9(2 \cos 7\theta) + 36(2\cos 5\theta) + 84 \cos (2 \cos 3\theta) + 126\,(2 \cos\theta)$$

$$\cos^9\theta = (1/2)^8\, [\cos 9\theta + 9 \cos 7\theta + 36 + 50 + 84 \cos 3\theta + 126 \cos \theta]$$

Ans.

Example 3:

Expand $\cos^8 \theta$.

Solution:

With the same notations as in we get $(2 \cos \theta)^8 = \left(x+\frac{1}{x}\right)^8$

$$= x^8 + {}^8c_1\, x^7 (1/x) + {}^8c_2 (1/x^2) + {}^8c_3 x^5 (1/x^3)$$
$$+ {}^8c_4\, x^4 (1/x^4) + {}^8c_5\, x^3 (1/x^5) + {}^8c_6 x^2 (1/x^6)$$
$$+ {}^3c_7\, x (1/x^7) (1/x^7) + {}^8c_8 (1/x)^8$$
$$= x^8 + 8x^6 + 28x^4 + 56x^2 + 70 + 56 (1/x^2)$$
$$+ 28 (1/x^4) + 8 (1/x^7) + {}^8c_8 (1/x)^8$$
$$= \left(x^8+\frac{1}{x^8}\right) + 8\left(x^6+\frac{1}{x^6}\right) + 28\left(x^4+\frac{1}{x^4}\right) + 56\left(x^2+\frac{1}{x^2}\right) + 70$$
$$\Rightarrow\ 2^8 \cos^8 \theta = (2 \cos 8\theta) + 8 (2 \cos 6\theta) + 28 (2 \cos 4\theta)$$
$$+ 56 (2 \cos 2\theta) + 70$$
$$\Rightarrow\ \cos^8 = \frac{1}{2^7} [\cos 8\theta + 8 \cos 6\theta + 28 \cos 4\theta + 56 \cos 2\theta + 35]$$ **Ans.**

EXPANSION OF SIN^N Θ IN A SERIES OF COSINES AND SINES OF MULTIPLES OF Θ, ACCORDING AS n (A POSITIVE INTEGER) IS EVEN OR ODD

We have $(2i \sin \theta)^n = (x - 1/x)^n$

$$\Rightarrow\ 2^n i^n \sin^n \theta = x^n - {}^nc_1\, x^{n-1} (1/x) + {}^nc_2\, x^{n-2} (1/x^2) - \dots$$
$$\dots + (-1)^n\, {}^nc_n (1/x^n)$$
$$\Rightarrow\ 2^n i^n \sin^n \theta = x^n - {}^nc_1\, x^{n-2} + {}^nc_2\, x^{n-4} - \dots$$
$$\dots + (-1)^n\, {}^nc_n (1/x^n) (1/x^n) \qquad \dots(i)$$

Case I n is even: In this case the number of terms in R.H.S. of (i) will be odd and hence the last term is positive, last but one is negative and so on.

Also the middle term which is independent of x is $(-1)^{n/2}\, {}^nc_{n/2}$

Also $i^n = (-1)^{n/2}$

Now as ${}^nc_r = {}^nc_{n-r}$, ${}^nc_{n-1} = {}^nc_1$, etc and ${}^nc_n = 1$,

∴ from (i) after taking the terms in pairs as in we have

$$2^n (-1)^{n/2} \sin^n \theta = \left(x^n+\frac{1}{x^n}\right) - {}^nc_1 \left(x^{n-2}+\frac{1}{x^{n-2}}\right)$$
$$+ {}^nc_2 \left(x^{n-4}+\frac{1}{x^{n-4}}\right) - \dots + (-1)^{n/2}\, {}^nc_{n/2}$$

$$\Rightarrow\ 2^n (-1)^{n/2} \sin^n \theta = (2 \cos n\theta) - {}^nc_1 [2 \cos (n-2) \theta] + {}^nc_2 [2 \cos (n-4) \theta] - \dots + (-1)^{n/2}\, {}^nc_{n/2}$$

$$\Rightarrow\ 2^{n-1} (-1)^{n/2} \sin^n \theta = \cos n\theta - {}^nc_1 \cos (n-2) \theta + {}^nc_2 \cos (n-4) \theta - \dots + \frac{(-1)^{n/2}}{2}\, {}^nc_{n/2}$$

Case II – n is odd: In this case the number of terms in R.H.S. of (i) will be even and hence the last term is negative, last but one is positive and so on. Also the two middle terms are

$$(-1)^{(n-1)/2}\, {}^nc_{(n-1)/2}\, x \text{ and } (-1)^{(n+1)/2}\, {}^nc_{(n+1)/2}\, \frac{1}{x}.$$

But ${}^nc_{(n-1)/2} = {}^nc_{(n+1)/2}$, hence these two middle terms taken in a pair = $(-1)^{(n-1)/2}\, {}^nc_{(n-1)/2}\, (x - 1/x)$.

Hence, in this case, from (i) after combining terms in pairs as in we have

$$2^n i^n \sin^n \theta = \left(x^n - \frac{1}{x^n}\right) - {}^nc_1 \left(x^{n-2} - \frac{1}{x^{n-2}}\right) + {}^nc_2 \left(x^{n-4} - \frac{1}{x^{n-4}}\right) + \dots + (-1)^{(n-1)/2}\, {}^nc_{(n-1)/2} \times (x - 1/x)$$

$$= (2i \sin n\theta) - {}^nc_2 [2i \sin (n-2) \theta] + {}^nc_4 [2i \text{ s n } (n-4) \theta] + \dots + (-1)^{(n-1)/2}\, {}^nc_{(n-1)/2} \times (2i \sin \theta)$$

$$\Rightarrow 2^{(n-1)/2}\, i^{n-1} \sin^n \theta = \sin n\theta - {}^nc_1 \sin (n-2) \theta + {}^nc_2 \sin (n-4) \theta - \dots + \frac{(-1)^{(n-1)/2}}{2}.{}^nc_{(n-1)/2} \sin \theta$$

$$\Rightarrow\ 2^{n-1} (-1)^{n-1/2} \sin^n \theta = \sin n\theta - {}^nc_1 \sin (n-2) \theta + {}^nc_2 \sin (n-4) \theta - \dots + \frac{(-1)^{(n-1)/2}}{2}\, {}^nc_{(n-1)/2} \sin \theta$$

Example 1:

Expand $\sin^7 \theta$.

Solution:

Let $x = \cos \theta + i \sin \theta$, then $(1/x) = \cos \theta - i \sin \theta$,

$x^n = \cos n\theta + i \sin n\theta$, $(1/x^n) = \cos n\theta - i \sin n\theta$,

whence $x - 1/x = 2i \sin \theta$ and $x^n - 1/x^n = 2i \sin n\theta$,

Now $(2i \sin \theta)^7 = \left(x - \dfrac{1}{x}\right)^7$

$$= x^7 - {}^7c_1\, x^6.\frac{1}{x} + {}^7c_2 x^5.\frac{1}{x^2} - {}^7c_3 x^4.\frac{1}{x^3} + {}^7c_4 x^3.\frac{1}{x^4}$$

$$- {}^7c_5 x^2.\frac{1}{x^5} + {}^7c_6 x.\frac{1}{x_6} - {}^7c_7 \frac{1}{x^7}$$

$$= x^7 - 7x^5 + 21x^3 - 35x + \frac{35}{x} - \frac{21}{x^3} + \frac{7}{x^5} - \frac{1}{x^7}$$

$$\Rightarrow\ 2^7 i^7 \sin^7\theta = \left(x^7 - \frac{1}{x^7}\right) - 7\left(x^5 - \frac{1}{x^5}\right) + 21\left(x^3 - \frac{1}{x^3}\right) - 35\left(x - \frac{1}{x}\right)$$

$\Rightarrow\ -2^7 i \sin^7\theta = (2i \sin 7\theta) - 7\,(2i \sin 5\theta) + 21\,(2i \sin 3\theta) - 35\,(2i \sin\theta)$

$$\Rightarrow\ -2^6 \sin^7\theta = \sin 7\theta - 7\sin 5\theta + 21 \sin 3\theta - 35 \sin\theta$$

$$\Rightarrow\ \sin^7\theta = -\frac{1}{2^6}\,[\sin 7\theta - 7\sin 5\theta + 21\sin 3\theta - 35\sin\theta]$$ **Ans.**

Example 2:

Expand $\sin^8\theta$.

Solution:

With the same notations as in above we have

$x + 1/x = 2\cos\theta$; $x - 1/x = 2i\sin\theta$; $x^n + 1/x^n = 2\cos n\theta$

Now $(2i\sin\theta)^8 = (x - 1/x)^8$

$$\Rightarrow\ 2^8 i^8 \sin^8\theta = x^8 - {}^8c_1\, x^7\, \frac{1}{x} + {}^8c_2\, x^5\, \frac{1}{x^3}$$

$$+ {}^8c_4 x^4.\frac{1}{x^4} - {}^8c_5 x^3\, x^3\, \frac{1}{x^5} + {}^8c_6 x^2.\frac{1}{x^6} - {}^8c_7\, x\, \frac{1}{x^7} + {}^8c_8\, \frac{1}{x^8}$$

$$= c^8 - 8x^6 + 28x^4 - 56x^2 + 70 - \frac{56}{x^2} + \frac{28}{x^4} - \frac{8}{x^6} + \frac{1}{x^8}$$

$$\Rightarrow\ 2^8 \sin^8\theta = \left(x^8 + \frac{1}{x^8}\right) - 8\left(x^6 + \frac{1}{x^6}\right) + 28\left(x^4 + \frac{1}{x^4}\right) 56\left(x^2 + \frac{1}{x^2}\right) + 70$$

$$= (2\cos 8\theta) - 8\,(2\cos 6\theta) + 28\,(2\cos 4\theta) - 56\,(2\cos 2\theta) + 70$$

$$\Rightarrow\ \sin^8\theta = \frac{1}{2^7}\,[\cos 8\theta - 8\cos 6\theta + 28\cos 4\theta - 56\cos 2\theta + 35]$$ **Ans.**

Example 3:

Expand $\sin^9\theta$.

Solution:

With the same notations as in above we have

$$(2i \sin \theta)^9 = (x - 1/x)^9$$

$$\Rightarrow \quad 2^9 i^9 \sin^9 \theta = x^9 - {}^9c_1 x^8 . \frac{1}{x} + {}^9c_2\, x^7 . \frac{1}{x^2} - {}^9c_3\, x^6 . \frac{1}{x^3}$$

$$+ {}^9c_4 x^5 . \frac{1}{x^4} - {}^9c_5 x^4 . \frac{1}{x^5} + {}^9c_6 x^3 . \frac{1}{x^6} - {}^9c_7\, x^2 . \frac{1}{x^7}$$

$$+ {}^9c_8\, x . \frac{1}{x^8} - {}^9c_9\, \frac{1}{x^9} .$$

$$\Rightarrow \quad 2^9 i \sin^9 \theta = x^9 - 9x^7 + 36x^5 - 84x^3 + 126x - \frac{126}{x} + \frac{84}{x^3} - \frac{36}{x^5} + \frac{9}{x^7} - \frac{1}{x^9} .$$

$$= \left(x^9 - \frac{1}{x^9}\right) - 9\left(x^7 - \frac{1}{x^7}\right) + 36\left(x^5 - \frac{1}{x^5}\right)$$

$$-84\left(x^3 - \frac{1}{x^3}\right) - 126\left(x - \frac{1}{x}\right)$$

$$\Rightarrow \quad 2^9\, i \sin^9 \theta = (2i \sin 9\theta) - 9\,(2i \sin 7\theta) + 36\,(2i \sin 5\theta)$$

$$- 84\,(2i \sin 3\theta) + 126\,(2i \sin \theta).$$

$$\Rightarrow \quad \sin^9 \theta = \frac{1}{2^9}\,[\sin 9\theta - 9 \sin 7\theta + 36 \sin 5\theta - 84 \sin 3\theta + 126 \sin\theta].$$

Ans.

Example 4:

Express $\cos^5 \theta \sin^3 \theta$ in terms of since of multiple of θ.

Solution:

Let $x = \cos \theta + i \sin \theta$, then $1/x = \cos \theta - i \sin \theta$,

$x^n = \cos n\theta + i \sin n\theta$ and $1/x^n = \cos n\theta - i \sin n\theta$,

whence $x + 1/x = 2 \cos \theta$, $x - 1/x = 2i \sin \theta$ etc.

Now $(2 \cos \theta)^5 (2i \sin \theta)^3 = (x + 1/x)^5 (x - 1/x)^3$ ...(i)

If we actually expand and multiply R.H.S. of (i), then the process becomes very lengthy and laborious. In order to avoid the same the following method is used to find out the coefficients of the products. **(Note)**

Coefficients of $\left(x + \frac{1}{x}\right)^5$

1 5 10 10 5 1

Multiplying by $\left(x - \frac{1}{x}\right)$, once, the coefficients are

1 (5 – 1) (10 – 5) (10 – 10) (5 – 10) (1 – 5) (0 – 1) **(Note)**

i.e., 1 4 5 0 – 5 – 4 – 1

Multiplying twice, coefficients are

1 3 1 –5 – 5 1 3 –1

Multiplying thrice, coefficients are

1 2 –2 – 6 0 6 2 – 1 – 1

Hence from (i) we have

$$(2\cos\theta)^5 (2i\sin\theta)^3 = x^8 + 2x^6 - 2x^4 - 6x^2 + 0 + \frac{6}{x^2} + \frac{2}{x^4} - \frac{2}{x^6} - \frac{1}{x^8}$$

$$\Rightarrow 2^8 i^3 \cos^5\theta \sin^3\theta = \left(x^8 - \frac{1}{x^8}\right) + 2\left(x^6 - \frac{1}{x^6}\right) - 2\left(x^4 - \frac{1}{x^4}\right) - 6\left(x^2 - \frac{1}{x^2}\right)$$

$$\Rightarrow -2^8 i \cos^5\theta \sin^3\theta = (2i\sin 8\theta) + 2(2i\sin 6\theta) - 2(2i\sin 4\theta) - 6(2i\sin 2\theta)$$

$$\Rightarrow -2^7 \cos^5\theta \sin^3\theta = \sin 8\theta + 2\sin 6\theta - 2\sin 4\theta - 6\sin 2\theta$$

$$\Rightarrow \cos^5\theta \sin^3\theta = -\frac{1}{128}[\sin 8\theta + 2\sin 6\theta - 2\sin 4\theta - 6\sin 2\theta].$$

Ans.

Example 5:

Prove that

$$32\sin^4\theta\cos^2\theta = \cos 6\theta - 2\cos 4\theta - \cos 2\theta + 2$$

Solution:

With the same notations as in above, we have

$$(2i\sin\theta)^4 (2\cos\theta)^2 = \left(x - \frac{1}{x}\right)^4 \left(x + \frac{1}{x}\right)^2 \quad \text{...(i)}$$

In order to find the coefficients of the product in R.H.S. of (i) we have as follows:

Coefficients of $\left(x - \frac{1}{x}\right)$ with proper sign are

1 – 4 6 – 4 1

Multiplying by $\left(x + \frac{1}{x}\right)$ once, the coefficients are

1 (– 4 + 1) (6 – 4) (– 4 + 6) (1 – 4) (0 + 1) **(Note)**

i.e., 1   – 3   2   2   – 3   1

Multiplying by $\left(x+\frac{1}{x}\right)$ twice, coefficients are

1   – 2   – 1   4   – 1   – 2   1

Hence from (i) we have

$$2^4\, i^4 \sin^4\theta\; 2^2 \cos^2\theta = x^6 - 2x^4 - x^2 + 4 - \frac{1}{x^2} - \frac{2}{x^4} + \frac{1}{x^6}$$

$$\Rightarrow\; 2^6 \sin^4\theta \cos^2\theta = \left(x^6+\frac{1}{x^6}\right) - 2\left(x^4+\frac{1}{x^4}\right) - \left(x^2+\frac{1}{x^2}\right) + 4$$

$$\Rightarrow\; 64 \sin^4\theta \cos^2\theta = (2\cos 6\theta) - 2\,(2\cos 4\theta) - (2\cos 2\theta) + 4$$

$\Rightarrow\; 32 \sin^4\theta \cos^2\theta = \cos 6\theta - 2\cos 4\theta - \cos 2\theta + 2.$ **Hence proved.**

Example 6:

Prove that

$-256 \sin^7\theta \cos^2\theta = \sin 9\theta - 5\sin 7\theta + 8\sin 5\theta - 14\sin\theta.$

Solution:

With the same notations as in we have

$$(2i\sin\theta)^7\,(2\cos\theta)^2 = \left(x-\frac{1}{x}\right)^7\left(x+\frac{1}{x}\right)^2 \qquad ...(i)$$

In order to find the coefficients of the product in R.H.S. of (i) we proceed as follows:

Coefficients of $\left(x-\frac{1}{x}\right)$ with proper sign are

1   – 7   21   – 35   35   – 21   7   –1

Multiplying by $\left(x+\frac{1}{x}\right)$ the coefficients are

1   – 6   14   – 14   0   14   – 14   6   –1

Multiplying by $\left(x+\frac{1}{x}\right)$ twice, the coefficients are

1   –5   8   0   – 14   14   0   – 8   5   –1

Hence from (i) we have

$$2^7\, i^7 \sin^7\theta\; 2^2 \cos^2\theta = x^9 - 5x^7 + 8x^5 + 0x^3 - 14x + \frac{14}{x} + \frac{0}{x^3}$$

$$-\frac{8}{x^5}+\frac{5}{x^7}-\frac{1}{x^9}$$

$$\Rightarrow \quad -2^9 \text{ i } \sin^7\theta \cos^2\theta = \left(x^9-\frac{1}{x^9}\right)-5\left(x^7-\frac{1}{x^7}\right)$$

$$+8\left(x^5-\frac{1}{x^5}\right)-14\left(x-\frac{1}{x}\right)$$

$$= (2i \sin 9\theta) - 5\,(2i \sin 7\theta) + 8\,(2i \sin 5\theta)$$

$$- 14\,(2i \sin \theta)$$

$$\Rightarrow \quad -2^8 \sin^7\theta \cos^2\theta = \sin 9\theta - 5 \sin 7\theta + 8 \sin 5\theta - 14 \sin\theta$$

$$\Rightarrow \quad -256 \sin^7\theta \cos^2\theta = \sin 9\theta - 5 \sin 7\theta + 8 \sin 5\theta - 14 \sin\theta.$$

Hence proved.

Example 7:

Express $\sin^6\theta \cos^2\theta$ in terms of cosines of multiples of θ.

Solution:

With the same notations as in we have

$$(2i \sin\theta)^6 (2\cos\theta)^2 = \left(x-\frac{1}{x}\right)^6\left(x+\frac{1}{x}\right)^2 \qquad ..(i)$$

In order to find the coefficients of the product in R.H.S. of (i) we proceed as follows:

Coefficients of $\left(x-\frac{1}{x}\right)^6$ with proper sign are

1   – 6   15   – 20   15   – 6   1

Multiplying by $\left(x+\frac{1}{x}\right)$ the coefficients are

1   – 5   9   – 5   – 5   9   – 5   1

Multiplying by $\left(x+\frac{1}{x}\right)$ twice, the coefficients are

1   – 4   4   4   – 10   4   4   – 4   1

Hence from (i) we have

$$2^6 i^6 \sin^6\theta\, 2^2 \cos^2\theta = x^8 - 4x^6 + 4x^4 + 4x^2 - 10 + \frac{4}{x^2}+\frac{4}{x^4}-\frac{4}{x^6}+\frac{1}{x^8}$$

$$\Rightarrow \quad -2^8 \sin^6\theta \cos^2\theta = \left(x^8+\frac{1}{x^8}\right)-4\left(x^6+\frac{1}{x^6}\right)$$

$$+4\left(x^4+\frac{1}{x^4}\right)+4\left(x^2+\frac{1}{x^2}\right)-10$$

$$\Rightarrow\ -2^8 \sin^6\theta \cos^2\theta = (2\cos 8\theta) - 4\,(2\cos 6\theta)$$
$$+ 4\,(2\cos 4\theta) + 4\,(2\cos 2\theta) - 10$$

$$\Rightarrow\ -2^7 \sin^6\theta \cos^2\theta = \cos 8\theta - 4\cos 6\theta + 4\cos 4\theta + 4\cos 2\theta - 5$$

$$\Rightarrow \sin^6\theta \cos^2\theta = -\frac{1}{2^7}[\cos 8\theta - 4\cos 6\theta + 4\cos 4\theta + 4\cos 2\theta - 5].$$

Ans.

Example 8:

Express $\cos^5\theta \sin^7\theta$ *in a series of sines of multiple of* θ.

Solution:

With the previous notations, we have

$$(2i\sin\theta)^7 (2\cos\theta)^5 = \left(x-\frac{1}{x}\right)\left(x+\frac{1}{x}\right)^5 \qquad \text{...(i)}$$

In order to find the coefficients of the product in R.H.S. of (i), we proceed as follows

Coefficients of $\left(x-\frac{1}{x}\right)$ with proper sign are

1   – 7   21   – 35   35   – 21   7   – 1

Multiplying by $\left(x+\frac{1}{x}\right)$, the coefficients are

1   – 6   14   –14   0   14   – 14   6   –1

Multiplying by $\left(x+\frac{1}{x}\right)$ twice, coefficients are

1   – 5   8   0   – 14   0   – 8   5   –1

Multiplying by $\left(x+\frac{1}{x}\right)$ thrice, coefficients are

1   – 4   3   8   – 14   0   14   – 8   – 3   4   – 1

Multiplying by $\left(x+\frac{1}{x}\right)$ four times, coefficients are

1   – 3   –1   11   – 6   – 14   6   11   1   3   – 1

Multiplying by $\left(x+\frac{1}{x}\right)$ five times, coefficients are

1 – 2 – 4 10 5 – 20 0 20 – 5 – 10 4 2 –1

Hence from (i) we have

$$2^7 i^7 \sin^7 \theta\, 2^5 \cos^5 \theta = x^{12} - 4x^8 + 10x^6 + 5x^4 - 20x^2 + 0 + \frac{20}{x^2} - \frac{5}{x^4} - \frac{10}{x^6} + \frac{4}{x^8} + \frac{2}{x^{10}} - \frac{1}{x^{12}}$$

$$\Rightarrow\ -2^{12} i \sin^7 \theta \cos^5 \theta = \left(x^{12} - \frac{1}{x^{12}}\right) - \left(x^{10} - \frac{1}{x^{10}}\right)$$

$$-4\left(x^8 - \frac{1}{x^8}\right) + 10\left(x^6 - \frac{1}{x^6}\right) + 5\left(x^4 - \frac{1}{x^4}\right) - 20\left(x^2 - \frac{1}{x^2}\right)$$

$$= (2i \sin 12\theta) - 2\,(2i \sin 10\theta) - 4\,(2i \sin 8\theta) + 10\,(2i \sin 6\theta) + 5\,(2i \sin 4\theta) - 20\,(2i \sin 2\theta)$$

$$\Rightarrow\ -2^{11} \sin^7 \theta \cos^5 \theta = \sin 12\theta - 2 \sin 10\theta - 4 \sin 8\theta + 10 \sin 6\theta + 5 \sin 4\theta - 20 \sin 2\theta$$

$$\Rightarrow\ \sin^7 \theta \cos^5 \theta = -\frac{1}{2^{11}}\,[\sin 12\theta - 2 \sin 10\theta - 4 \sin 8\theta + 10 \sin 6\theta + 5 \sin 4\theta - 20 \sin 2\theta]$$

Ans.

EXPANSIONS OF COS^N Θ AND SIN^N Θ IN TERMS OF SINES AND COSINES OF MULTIPLES OF Θ, N BEING A POSITIVE INTEGER

Let $z = \cos \theta + i \sin \theta$ so that $\frac{1}{z} = \cos \theta - i \sin \theta$.

Now $z + \frac{1}{z} = 2 \cos \theta$ and $z - \frac{1}{z} = 2i \sin \theta$.

By Expansion of Some Trigonometrical Functions, we have

$z^n = \cos n\theta + i \sin n\theta$ and $\frac{1}{z^n} = \cos n\theta - i \sin n\theta$.

Thus $z^n + \frac{1}{z^n} = 2 \cos n\theta$ and $z^n - \frac{1}{z^n} = 2i \sin n\theta$. ...(2)

From (2), we have

$$(2 \cos \theta)^n = \left(z + \frac{1}{z}\right)^n.$$

Expanding the R.H.S. with the help of the Binomial Theorem and using (2), we get the required expansion of $\cos^n \theta$. Similarly

$$(2i \sin \theta)^n = \left(z - \frac{1}{z}\right)^n$$

gives the required expansion of $\sin^n \theta$.

Example 1:

Prove that

(i) $16 \sin^5 \theta = \sin 5\theta - 5 \sin 3\theta + 10 \sin \theta$.

(ii) $16 \cos^5 \theta = \cos 5\theta + 5 \cos 3\theta + 10 \cos \theta$.

Solution:

Let $z = \cos \theta + i \sin \theta$ so that $1/z = \cos \theta - i \sin \theta$.

$$\Rightarrow \quad z - \frac{1}{z} = 2i \sin \theta$$

$$\Rightarrow \quad (2i \sin \theta)^5 = \left(z - \frac{1}{z}\right)^5$$

$$= z^5 - {}^5C_1\, z^4 . \frac{1}{z} + {}^5C_2\, z^3 . \frac{1}{z^2} - {}^5C_3 z^2 . \frac{1}{z^3} + {}^5C_4 z . \frac{1}{z^4} - \frac{1}{z^5}$$

$$= z^5 - 5\, z^3 + \frac{5.4}{2.1} z - \frac{5.4.3}{3.2.1} . \frac{1}{z} + \frac{5.4.3.2}{4.3.2.1} . \frac{1}{z^3} - \frac{1}{z^5}$$

$$= \left(z^5 - \frac{1}{z^5}\right) - 5\left(z^3 - \frac{1}{z^3}\right) + 10\left(z - \frac{1}{z}\right).$$

$$= (2i \sin 5\theta) - 5\,(2i \sin 3\theta) + 10\,(2i \sin \theta), \text{ by (2) above}$$

$$\Rightarrow \quad 32i \sin^5 \theta = 2i\,(\sin 5\theta - 5 \sin 3\theta + 10 \sin \theta) \qquad (\because\ i^5 = i)$$

Hence $16 \sin^5 \theta = \sin 5\theta - 5 \sin 3\theta + 10 \sin \theta$.

(ii) We have $z + 1/z = 2 \cos \theta$, and so

$$(2 \cos \theta)^5 = \left(z + \frac{1}{z}\right)^5 = z^5 + 5z^3 + 10z + 10\, \frac{1}{z} + 5\, \frac{1}{z^3} + \frac{1}{z^5},$$

(as shown in the first part)

$$= \left(z^5 + \frac{1}{z^5}\right) + 5\left(z^3 + \frac{1}{z^3}\right) + 10\left(z + \frac{1}{z}\right)$$

$$\Rightarrow \quad 2^5 \cos^5 \theta = (2 \cos 5\theta) + 5\,(2 \cos 3\theta) + 10\,(2 \cos \theta), \text{ by (2) above.}$$

Hence $16 \cos^5 \theta = \cos 5\theta + 5 \cos 3\theta + 10 \cos \theta$.

Example 2:

Prove that $32 \cos^6 \theta = \cos 6\theta + 6 \cos 4\theta + 15 \cos 2\theta + 10$.

Solution:

Let $x = \cos \theta + i \sin \theta$ so that $1/x = \cos \theta - i \sin \theta$.

Thus $2\cos\theta = x + \frac{1}{x} \Rightarrow (2\cos\theta)^6 = \left(x+\frac{1}{x}\right)^6$

$$= x^6 + {}^6C_1x^5 \cdot \frac{1}{x} + {}^6C_2x^4 \cdot \frac{1}{x^2} + {}^6C_3x^3 \cdot \frac{1}{x^3} + {}^6C_4x^2 \cdot \frac{1}{x^4} + {}^6C_5x \cdot \frac{1}{x^5} + \frac{1}{x^6}$$

$$= x^6 + 6x^4 + 15x^2 + + 20 + \frac{15}{x^2} + \frac{6}{x} + \frac{1}{x^6}$$

$$= \left(x^6 + \frac{1}{x^6}\right) + 6\left(x^4 + \frac{1}{x^4}\right) + 15\left(x^2 + \frac{1}{x^2}\right) + 20$$

$\therefore\ 64\cos^6\theta = 2\cos 6\theta + 6.2\cos 4\theta + 15.2\cos 2\theta + 20$

Hence $32\cos^6\theta = \cos 6\theta + 6\cos 4\theta + 15\cos 2\theta + 10$.

Example 3:

Prove that

$$64\cos^7\theta = \cos 7\theta + 7\cos 5\theta + 21\cos 3\theta + 35\cos\theta.$$

Solution:

Let $x = \cos\theta + i\sin\theta$, then $1/x = \cos\theta - i\sin\theta$.

$$\therefore\quad (2\cos\theta)^7 = \left(x+\frac{1}{x}\right)^7 = x^7 + 7c_1.x^6 \cdot \frac{1}{x} + 7\,c_2.x^5 \cdot \frac{1}{x^2}$$

$$+ 7c_3x^4.\frac{1}{x^3} + 7c_4.x^3.\frac{1}{x^4} + 7c_5x^2. + 7c_6x. + 7c_7.\frac{1}{x^7}$$

$$= x^7 + 7x^5 + 21x^3 + 35x + 35.\frac{1}{x} + 21.\frac{1}{x^3} + 7.\frac{1}{x^5} + \frac{1}{x^7}$$

$$= \left(x^7 + \frac{1}{x^7}\right) + 7\left(x^5 + \frac{1}{x^5}\right) + 21\left(x^3 + \frac{1}{x^3}\right) + 35\left(x + \frac{1}{x}\right)$$

$= 2\,[\cos 7\theta + 7\cos 5\theta + 21\cos 3\theta + 35\cos\theta]$.

Hence $64\cos^7\theta = \cos 7\theta + 7\cos 5\theta + 21\cos 3\theta + 35\cos\theta$.

Example 4:

Prove that

$$\sin^8\theta = \frac{1}{128}\{\cos 8\theta - 8\cos 6\theta + 28\cos 4\theta - 56\cos 2\theta + 70\}.$$

Solution:

Let $z = \cos\theta + i\sin\theta \Rightarrow z^{-1} = \cos\theta - i\sin\theta$

$$\Rightarrow \quad z - z^{-1} = 2\,i \sin\theta \Rightarrow \sin^8\theta = \frac{1}{(2i)^8}\,(z - z^{-1})^8$$

$$= \frac{1}{256}\,(z^8 - {}^8c_1\, z^6 + {}^8c_2\, z^4 - {}^8c_3\, z^2 + {}^8c_4 - {}^8c_5\, z^{-2}$$

$$+ {}^8c_6\, z^{-4} - {}^8c_7\, z^{-6} + z^{-8})$$

$$= \frac{1}{256}\,\{(z^8 + z^{-8}) - 8\,(z^6 + z^{-6}) + 28\,(z^4 + z^{-4}) - 56\,(z^2 + z^{-2}) +$$

70}

$$= \frac{1}{128}\,\{\cos 8\theta - 8 \cos 6\theta + 28 \cos 4\theta - 56 \cos 2\theta + 70\}.$$

Example 5:

Show that

$$\sin^6\theta = -\frac{1}{32}\,[\cos 6\theta - 6 \cos 4\theta + 15 \cos 2\theta - 10].$$

Solution:

We have $(2i \sin\theta)^6 = \left(x - \frac{1}{x}\right)^6$

$$= x^6 + 6c_1x^5\left(-\frac{1}{x}\right) + 6c_2x^4\left(-\frac{1}{x}\right)^2$$

$$+ 6c_3x^3\left(-\frac{1}{x}\right)^3 + 6c_4x^2\left(-\frac{1}{x}\right)^4 + 6c_5x^2\left(-\frac{1}{x}\right)^5 + 6c_6\left(-\frac{1}{x}\right)^6$$

$$= x^6 - 6x^4 + 15x^2 - 20 + 15\,\frac{1}{x^2} - 6\,\frac{1}{x^4} + \frac{1}{x^6}$$

$$= \left(x^6 + \frac{1}{x^6}\right) - 6\left(x^4 + \frac{1}{x^4}\right) + 15\left(x^2 + \frac{1}{x^2}\right) - 20$$

$$= [2 \cos 6\theta] - 6\,[2 \cos 4\theta] + 15\,[2 \cos 2\theta] - 20$$

Hence $\sin^6\theta = -\frac{1}{32}\,[\cos 6\theta - 6 \cos 4\theta + 15 \cos 2\theta - 10]$

Example 6:

Prove that: $32 \sin^4\theta \cos^2\theta = \cos 6\theta - 2 \cos 4\theta - \cos 2\theta + 2.$

Solution:

Let $x = \cos\theta + i \sin\theta \Rightarrow 1/x = \cos\theta - i \sin\theta.$

$\Rightarrow \quad x + 1/x = 2 \cos\theta$ and $x - 1/x = 2i \sin\theta,$

$$\Rightarrow \quad (2i \sin \theta)^4 (2 \cos \theta)^2 = \left(x+\frac{1}{x}\right)^2 \left(x-\frac{1}{x}\right)^4$$

$$= \left(x+\frac{1}{x}\right)^3 \left(x-\frac{1}{x}\right)^2 \left(x-\frac{1}{x}\right)^2$$

$$= \left(x^2-\frac{1}{x^2}\right)^2 \left(x-\frac{1}{x}\right)^2$$

$$= \left(x^4-2+\frac{1}{x^4}\right)\left(x^2-2+\frac{1}{x^2}\right)$$

$$= x^6 - 2x^4 - x^2 + 4 - \frac{1}{x^2} - \frac{2}{x^4} + \frac{1}{x^6}$$

$$= \left(x^6+\frac{1}{x^6}\right) - 2\left(x^4+\frac{1}{x^4}\right) - \left(x^2+\frac{1}{x^2}\right) + 4$$

$$= 2 \cos 6\theta - 2.2 \cos 4\theta - 2 \cos 2\theta + 4.$$

Hence $32 \sin^4 \theta \cos^2 \theta = \cos 6\theta - 2 \cos 4\theta - \cos 2\theta + 2.$

Example 7:

Prove that $128 \cos^3 \theta \sin^5 \theta = \sin 8\theta - 2 \sin 6\theta - 2 \sin 4\theta + 6 \sin 2\theta.$

Solution:

If $z = \cos \theta + i \sin \theta$, then $z^{-1} = \cos \theta - i \sin \theta.$

$\Rightarrow \quad 2 \cos \theta = z + z^{-1},\ 2i \sin \theta = z - z^{-1}.$

$\Rightarrow \quad (2 \cos \theta)^3 (2i \sin \theta)^5 = (z + z^{-1})^3 (z - z^{-1})^5$

$= (z^2 - z^{-2})^3 (z - z^{-1})^2$

$= (z^6 - 3z^2 + 3z^{-2} - z^{-6}) (z^2 - 2 + z^{-2})$

$= (z^8 - z^{-8}) - 2 (z^6 - z^{-6}) - 2 (z^4 - z^{-4}) + 6 (z^2 - z^{-2})$

$2^8 i \cos^3 \theta \sin^5\theta = 2i \sin 8\theta - 2(2i \sin 6\theta) - 2(2i \sin 4\theta) + 6(2i \sin 2\theta)$

Hence $128 \cos^3 \theta \sin^5 \theta = \sin 8\theta - 2 \sin 6\theta - 2 \sin 4\theta + 6 \sin 2\theta.$

Example 8:

Prove that

$64 \sin^3 \theta \cos^4 \theta = -\sin 7\theta - \sin 5\theta + 3 \sin 3\theta + 3 \sin \theta.$

Solution:

$(2 \cos \theta)^4 (2i \sin \theta)^3 = (z + z^{-1})^4 (z - z^{-1})^3$

$$= (z^4 + 4z^2 + 6 + 4z^{-2} + z^{-4})(z^3 - 3z + 3z^{-1} - z^{-3})$$

$$= (z^7 - z^{-7}) + (z^5 - z^{-5}) - 3(z^3 - z^{-3}) - 3(z - z^{-1})$$

$$\Rightarrow \; -2i(64 \sin^3\theta \cos^4\theta) = 2i \sin 7\theta + 2i \sin 5\theta - 3(2i \sin 3\theta) - 3(2i \sin\theta)$$

Hence $64 \sin^2\theta \cos^4\theta = -\sin 7\theta - \sin 5\theta + 3 \sin 3\theta + 3 \sin\theta$.

Example 9:

Expand $\sin^7\theta \cos^2\theta$ in a series of sines of multiplies of θ.

Solution:

Let $z = \cos\theta + i \sin\theta$.

$$\therefore \quad (2i \sin\theta)^7 (2\cos\theta)^3 = \left(z - \frac{1}{z}\right)^7 \left(z + \frac{1}{z}\right)^3$$

$$= \left(z - \frac{1}{z}\right)^4 \left(z^2 - \frac{1}{z^2}\right)^3$$

$$= (z^4 - 4z^2 + 6 - 4z^{-2} + z^{-4})(z^6 - 3z^2 + 3z^{-2} - z^{-6})$$

$$= z^{10} - 4z^8 + 3z^6 + 8z^4 - 14z^2 + 14z^{-8} - 8z^{-4} - 3z^{-6} + 4z^{-8} - z^{-10}$$

$$= (z^{10} - z^{-10}) - 4(z^8 - z^{-8}) + 3(z^6 - z^{-6}) + 8(z^4 - z^{-4}) - 14(z^2 - z^{-2})$$

$$= 2i[\sin 10\theta - 4 \sin 8\theta + 3 \sin 6\theta + 8 \sin 4\theta - 14 \sin 2\theta].$$

Hence $-2^9 \sin^7\theta \cos^2\theta = \sin 10\theta - 4 \sin 8\theta + 3 \sin 6\theta + 8 \sin 4\theta - 14 \sin 2\theta$.

Example 10:

Prove that

$-128 \sin^6\theta \cos^2\theta = \cos 8\theta - 4 \cos 6\theta + 4 \cos 4\theta + 4 \cos 2\theta - 5$.

Solution:

Let $z = \cos\theta + i \sin\theta$ so that $1/z = \cos\theta - i \sin\theta$.

Thus $z + 1/z = 2\cos\theta$ and $z - 1/z = 2i \sin\theta$.

$$\therefore \quad (2i \sin\theta)^6 (2\cos\theta)^2 = \left(z - \frac{1}{z}\right)^6 \left(z - \frac{1}{z}\right)^2$$

$$= \left(z - \frac{1}{z}\right)^4 \left(z^2 - \frac{1}{z^2}\right)^2$$

$$= \left(z^4 - 4z^8 . \frac{1}{z} + 6z^2 . \frac{1}{z^2} - 4z . \frac{1}{z^3} + \frac{1}{z^4}\right)\left(z^4 - 2 + \frac{1}{z^4}\right)$$

$$= \left(z^8 + \frac{1}{z^8}\right) - 4\left(z^6 + \frac{1}{z^6}\right) + 4\left(z^4 + \frac{1}{z^4}\right) + 4\left(z^2 + \frac{1}{z^2}\right) + 10$$

$$= 2 \cos 8\theta - 4 (2 \cos 6\theta) + 4 (2 \cos 4\theta) + 4 (2 \cos 2\theta) - 10$$

$$\Rightarrow - 2^8 \sin^6 \cos^2 \theta = 2 (\cos 8\theta - 4 \cos 6\theta + 4 \cos 4\theta + 4 \cos 2\theta - 5)$$

Hence $- 128 \sin^6 \theta \cos^2 \theta = \cos 8\theta - 4 \cos 6\theta + 4 \cos 4\theta + 4\cos 2\theta - 5.$

Example 11:

Prove that

$128 \sin^2 \theta \cos^6 \theta = - 1 \cos 8\theta - 4 \cos 6\theta - 4 \cos 4\theta + 4 \cos 2\theta + 5.$

Solution:

Let $z = \cos \theta + i \sin \theta$ so that $z^{-1} = \cos \theta - i \sin \theta$.

$\therefore \quad z + z^{-1} = 2 \cos \theta, z - z^{-1} = 2i \sin \theta.$

We have

$$(2i \sin \theta)^2 (2 \cos \theta)^6 = (z - z^{-1})^2 (z + z^{-1})^6$$

$$= (z - z^{-1})^2 (z + z^{-1})^2 (z + z^{-1})^4$$

$$= (z^2 - z^{-2})^2 (z^4 + 4c_1 z^3 z^{-1} + 4c_3 z^2 z^{-2} + 4c_3 z z^{-3} + z^{-4})$$

$$= (z^4 - 2 + z^{-4}) (z^4 + 4z^2 + 6 + 4z^{-2} + z^{-4})$$

$$= (z^8 + z^{-8}) + 4 (z^6 + z^{-6}) + 4 (z^4 + z^{-4}) - 4 (z^2 + z^{-2}) - 10$$

$$\therefore \quad - 2^8 \sin^2 \theta \cos^6 \theta = 2 \cos 8\theta + 4 (2 \cos 6\theta)$$

$$+ 4 (2 \cos 4\theta) - 4 (2 \cos 2\theta) - 10.$$

Hence

$128 \sin^2 \theta \cos^6 \theta = - \cos 8\theta - 4 \cos 6\theta - 4 \cos 4\theta + 4 \cos 2\theta + 5.$

Example 12:

Prove that

$256 \sin^5 \theta \cos^4 \theta = \sin 9\theta - \sin 7\theta - 4 \sin 5\theta + 4 \sin 3\theta + 6 \sin \theta.$

Solution:

We have

$$(2i \sin \theta)^5 (2 \cos \theta)^4 = (z - z^{-1})^5 (z + z^{-1})^4$$

$$= (z^2 - z^{-2}) (z - z^{-1})$$

$$= (z^8 - 4z^4 + 6 - 4z^{-4} + z^{-8}) (z - z^{-1})$$

$$= (z^9 - z^{-9}) - (z^7 - z^{-7}) - 4 (z^5 - z^{-5}) + 4 (z^3 - z^{-3}) + 6 (z - z^{-1})$$

$\therefore$ i $2^9 \sin^5 \theta \cos^4 \theta = 2 i \sin 9\theta - 2i \sin 7\theta - 4 (2i \sin 5\theta)$
$+ 4 (2i \sin 3\theta) + (2i \sin \theta)$.

Hence

$256 \sin^5 \theta \cos^4 \theta = \sin 9\theta - \sin 7\theta - 4 \sin 5\theta + 4 \sin 3\theta + 6 \sin \theta$.

EXPRESSION FOR TAN ($\Theta_1 + \Theta_2 + ... + \Theta_N$) IN TERMS OF TAN Θ_1, TAN Θ_2,..., TAN Θ_N

We have $\cos (\theta_1 + \theta_2 +...+ \theta_n) + i \sin (\theta_1 + \theta_2 + ... + \theta_n)$

$= (\cos \theta_1 + i \sin \theta_1) (\cos \theta_2 + i \sin \theta_2) ... (\cos \theta_n + i \sin \theta_n)$

$= \cos \theta_2 \cos \theta_2 ... \cos \theta_n (1 + i \tan \theta_1) (1 + i \tan \theta_2).....$
$(1 + i \tan \theta_n)$

$= \cos \theta_1 \cos \theta_2 ... \cos \theta_n (1 + iS_1 - S_2 - iS_3 + S_4 + ...)$,
...(1)

where $S_1 = \tan \theta_1 + \tan \theta_2 + ... + \tan \theta_n = S \tan \theta_1$,

$S_2 = \tan \theta_1 \tan \theta_2 + ... = \Sigma \tan \theta_1 \tan \theta_2$,

$S_3 = \Sigma \tan \theta_1 \tan \theta_2 \tan \theta_3$, and so on.

Equating real and imaginary parts in (1), we have

$\cos (\theta_1 + \theta_2 + ... + \theta_n) = \cos \theta_1 \cos \theta_2 ... \cos \theta_n (1 - S_2 + S_4 ...)$,

and $\sin (\theta_1 + \theta_2 + ... + \theta_n) = \cos \theta_1 \cos \theta_2 ... \cos \theta_n (S_1 - S_3 + ...)$.

Hence $\tan (\theta_1 + \theta_2 + ... + \theta_n) = \dfrac{S_1 - S_3 + ...}{1 - S_2 + S_4 + ...}$.

Example 1:

If α, β, γ are the roots of $x^3 + px^3 + qx + p = 0$, prove that $\tan^{-1}\alpha + \tan^{-1}\beta + \tan^{-1}\gamma = n\pi$, except when $q = 1$.

Solution:

Since α, β, γ are the roots of the given equation,

$S_1 = \alpha + \beta + \gamma = -p$, $S_2 = \alpha\beta + \beta\gamma + \gamma\alpha = q$, $S_3 = \alpha\beta\gamma = -p$.

If $\alpha = \tan \theta_1$, $\beta = \tan \theta_2$, $\gamma = \tan \theta_3$, then

$S_1 = \Sigma \tan \theta_1$, $\beta = \tan \theta_2$, $\gamma = \tan \theta_3$, then

$S_1 = \Sigma \tan \theta_1$, $S_2 = \Sigma \tan \theta_1 \tan \theta_2$, $S_3 = \tan \theta_1 \tan \theta_2 \tan \theta_3$.

Now $\tan (\theta_1 + \theta_2 + \theta_3) = \dfrac{S_1 - S_3}{1 - S_2} = \dfrac{-p - (-p)}{1 - q} = \dfrac{0}{1 - q}$,

$\Rightarrow$ $\tan(\theta_1 + \theta_2 + \theta_3) = 0$, except when $q = 1$,

$\Rightarrow$ $\tan(\theta_1 + \theta_2 + \theta_3) = \tan n\pi$, except when $q = 1$,

$\Rightarrow$ $\theta_1 + \theta_2 + \theta_3 = np$, except when $q = 1$.

Hence $\tan^{-1}\alpha + \tan^{-1}\beta + \tan^{-1}\gamma = n\pi$, except when $q = 1$.

Example 2:

Prove that the equation $\cos 2\theta + p\cos\theta + q\sin\theta + r = 0$ has, in general, four roots between 0 and 2π, and that the sum of the four roots is a multiple of 2π.

Solution:

If $\tan\frac{\theta}{2} = t$, then $\cos\theta = \frac{1-t^2}{1+t^2}$, $\sin\theta = \frac{2t}{1+t^2}$.

Now $\cos 2\theta = 2\cos^2\theta - 1 = 2\left(\frac{1-t^2}{1+t^2}\right)^2 - 1 = \frac{1-6t^2+t^4}{(1+t^2)^2}$.

Substituting these values in the given equation, we get

$$\frac{1-6t^2+t^4}{(1+t^2)^2} + p\left(\frac{1-t^2}{1+t^2}\right) + q\left(\frac{2t}{1+t^2}\right) + r = 0$$

$\Rightarrow$ $(1 - 6t^2 + t^4) + p(1 - t^2)(1 + t^2) + 2qt(1 + t^2) + r(1 + t^2)^2 = 0$

$\Rightarrow$ $(1 - p + r)t^4$: $2qt^3 + 2(r - 3)t^2 + 2qt + (1 + p + r) = 0$.

Let the roots of this equation be t_1, t_2, t_3 and t_4, and

$t_4 = \tan(\theta_i/2)$, $i = 1, 2, 3, 4$.

Then $S_1 = \Sigma\tan\frac{\theta}{2} = \Sigma t_1 = -\frac{2q}{1-p+r}$

and $S_3 = \Sigma t_1t_2t_2 = -\frac{2q}{1-p+r}$. Also $1 - S_2 + S_4 \neq 0$.

Now $\tan\frac{1}{2}(\theta_1 + \theta_2 + \theta_3 + \theta_4) = \frac{S_1 - S_3}{1 - S_2 + S_4} = 0$,

$= \tan n\pi$, n being any integer.

$\Rightarrow$ $\frac{1}{2}(\theta_1 + \theta_2 + \theta_3 + \theta_4) = n\pi$. Hence $\theta_1 + \theta_2 + \theta_3 + \theta_4 = 2n\pi$.

Example 3:

If θ_1, θ_2, θ_3 be the three values of θ which satisfy the equation $\tan 2\theta = k\tan(\theta + \alpha)$ and such that no two of them differe by a multiple of π, show $\theta_1 + \theta_2 + \theta_3 + \alpha$ is a multiple of π.

Solution:

We have tan 2θ = k tan (θ + α)

$$\Rightarrow \quad \frac{2\tan\theta}{1-\tan^2\theta} = k.\frac{\tan\theta+\tan\alpha}{1-\tan\theta\tan\alpha}$$

$$\Rightarrow \quad 2\tan\theta\,(1-\tan\theta\tan\alpha) = k\,(\tan\theta+\tan\alpha)\,(1-\tan^2\theta)$$

$$\Rightarrow \quad 2\tan\theta - 2\tan^2\theta\tan\alpha$$

$$= k\,(\tan\theta - \tan^3\theta + \tan\alpha - \tan\alpha\tan^2\theta)$$

$$\Rightarrow \quad k\tan^3\theta + (k-2)\tan\alpha\tan^2\theta - (k-2)\tan\theta - k\tan\alpha = 0.$$

This equation being a cubic in tan θ has three roots. Let the three roots be $\tan\theta_1$, $\tan\theta_2$, $\tan\theta_3$.

$$\therefore \quad S_1 = \Sigma\tan\theta_1 = -\frac{k-2}{k}\tan\alpha.$$

$$S_2 = \Sigma\tan\theta_1\tan\theta_2 = -\frac{k-2}{k},$$

$$S_3 = \tan\theta_1\tan\theta_2\tan\theta_3 = \tan\alpha.$$

$$\text{Now } \tan(\theta_1+\theta_2+\theta_3) = \frac{S_1-S_3}{1-S_2} = \frac{\frac{1}{k}(2-k)\tan\alpha - \tan\alpha}{1+\frac{1}{k}(k-2)}$$

$$= \frac{2\tan\alpha - k\tan\alpha - k\tan\alpha}{k+k-2}$$

$$= \frac{2(k-1)\tan\alpha}{2(k-1)} = -\tan\alpha = \tan(n\pi-\alpha).$$

Thus $\theta_1 + \theta_2 + \theta_3 = n\pi - \alpha$, where n is any integer.

Hence $\theta_1 + \theta_2 + \theta_3 + \alpha = n\pi$, where n is any integer.

Example 4:

Prove that the equation

$$a^2\cos^2\theta + b^2\sin^2\theta + 2ga\cos\theta + 2fb\sin\theta + c = 0$$

has four roots and that the sum of the values of θ which satisfy it is an even multiple of π radians.

Solution:

Let $\tan\frac{\theta}{2} = t$. Then $\sin\theta = \frac{2t}{1+t^2}$, $\cos\theta = \frac{1-t^2}{1+t^2}$.

The given equation becomes

$$a^2\left(\frac{1-t^2}{1+t^2}\right)^2+b^2\left(\frac{2t}{1+t^2}\right)^2+2ga\left(\frac{1-t^2}{1+t^2}\right)+2fb\left(\frac{2t}{1+t^2}\right)+c=0$$

$\Rightarrow$ $a^2(1-t^2)^2+4b^2t^2+2ga(1-t^2)(1+t^2)+4fbt(1+t^2)^2$
$+ c(1+t^2)^2 = 0$

$\Rightarrow$ $t^4(c-2ga+a^2)+4fbt^2+t^2(2c+4b^2-2a^2)+4bft+$
$(c+2ga+a^2)=0.$

This is a fourth degree equation in t and so it has four roots say t_1, t_2, t_3, t_4. Let $t_i = \tan(\theta_i/2)$, i = 1, 2, 3, 4.

Now $S_1 = \Sigma t_1 = \Sigma \tan\frac{\theta_1}{2} = \frac{-4fb}{c-2ga+a^2}$,

$S_3 = \Sigma t_1t_2t_3 = \frac{-4fb}{c-2ga+a^2}$. Also $1 - S_2 + S_4 \neq 0$.

$\therefore \tan\frac{1}{2}(\theta_1+\theta_2+\theta_3+\theta_4) = \frac{S_1-S_3}{1-S_2+S_4} = 0 = \tan n\pi.$

$\Rightarrow \frac{1}{2}(\theta_1+\theta_2+\theta_3+\theta_4) = n\pi,$

Hence $\theta_1+\theta_2+\theta_3+\theta_4 = 2n\pi$, where n is any integer.

Example 5:

Show that the equation $\cos(2\theta-\alpha) = p\cos(\theta-\beta)$ can be satisfied by four values θ_1, θ_2, θ_3, θ_4 of θ, of which no two differe by a multiple of 2π. Show that $\theta_1+\theta_2+\theta_3+\theta_4-2\alpha = k\pi$, k being an even integer.

Solution:

We have $\cos(2\theta-\alpha) = p\cos(q-\beta)$

$\Rightarrow$ $\cos[(\theta-\alpha/2)+(\theta-\alpha/2)] = p[\cos(\theta-\alpha/2)+(\alpha/2-\beta)]$

$\Rightarrow$ $\cos 2\theta = p\cos(\phi+c)$, where $\phi = \theta-\alpha/2$ and $c = \alpha/2-\beta$

$\Rightarrow$ $\cos^2\phi - \sin^2\phi = p[\cos c\cos\phi - \sin c\sin\phi]$

$$\Rightarrow \left(\frac{1-t^2}{1+t^2}\right)-\left(\frac{2t}{1+t^2}\right)^2 = p\cos c\left(\frac{1-t^2}{1+t^2}\right)-p\sin c\left(\frac{2t}{1+t^2}\right),$$

$t = \tan\phi/2.$

$\Rightarrow$ $(1-t^2)^2-4t^2 = p\cos c(1-t^2)(1+t^2)-2p\sin ct(1+t^2)$

$\Rightarrow$ $1+t^4-6t^2 = p\cos c(1-t^4)-2p\sin c(1+t^3)$

$\Rightarrow$ $(1 + p \cos c)\, t^4 + 2p \sin c\, t^3 - 6t^2 + 2p \sin ct + (1 - p \cos c) = 0.$

This is a fourth degree equation in t and so it has four values of t say t_1, t_2, t_3 and t_4. We have $t_i = \tan(\phi_i/2)$, i = 1, 2, 3, 4.

$$\therefore \quad S_1 = \Sigma t_1 = \frac{-2p \sin c}{1 + p \cos c},$$

$$S_3 = \Sigma t_1 t_2 t_3 = \frac{-2p \sin c}{1 + p \cos c}, \text{ so that } S_1 - S_3 = 0.$$

$$\text{Now } \tan \frac{1}{2}(\phi_1 + \phi_2 + \phi_3 + \phi_4) = \frac{S_1 - S_3}{1 - S_1 + S_4} = 0, \ (1 - S_2 + S_4 \neq 0)$$

$$\therefore \quad \frac{1}{2}(\phi_1 + \phi_2 + \phi_3 + \phi_4) = n\pi, \text{ n being any integer.}$$

$\Rightarrow$ $\phi_1 + \phi_2 + \phi_3 + \phi_4 = 2n\pi$

$\Rightarrow$ $(\theta_1 - \alpha/2) + (\theta_2 - \alpha/2) + (\theta_3 - \alpha/2) + (\theta_4 - \alpha/2) = k\pi$, k = 2n

$(\therefore\ \phi = \theta - \alpha/2)$

Hence $\theta + \theta_2 + \theta_3 + \theta_4 - 2\alpha = k\pi$, k being an even integer.

Example 6:

Show that there are four values of θ lying between 0 and 2π which satisfy the equation ap sec θ – bq cosec $\theta = c^2$, and that their sum is an odd multiple of 2π.

Solution:

We know

$$\cos\theta = \frac{1 - t^2}{1 + t^2} \sin\theta = \frac{2t}{1 + t^2}, \text{ when } \tan\frac{\theta}{2} = t.$$

The equation ap sec θ – bq cosec $\theta = c^2$ becomes

$$\frac{ap\left(1 + t^2\right)}{1 - t^2} - \frac{bq\left(1 + t^2\right)}{2t} = c^2$$

$\Rightarrow$ $bqt^4 + t^2(2ap + 2c^2) = 0t^2 + t(2ap - 2c^2) - bq = 0.$

Since (1) is a fourth degree equation in t, so it has four values of t say t_1, t_2, t_3 and t_4. We have $t_i = \tan\theta/2$.

Then $S_1 = \Sigma t_1 = -(2ap + 2c^2)/bq$, $S_2 = \Sigma t_1 t_2$

$S_3 = \Sigma t_1 t_2 t_3 = -(2ap - 2c^2)/bq$, $S_4 = t_1 t_2 t_3 t_4 = -1$.

We know $\tan \frac{1}{2}(\theta_1+\theta_2+\theta_3+\theta_4) = \frac{S_1-S_3}{1-S_2+S_4}$

$$\Rightarrow \cot \frac{1}{2}(\theta_1+\theta_2+\theta_3+\theta_4) = \frac{1-S_2+S_4}{S_1-S_3} = 0, \ (S_1 - S_3 \neq 0)$$

$$\Rightarrow \frac{1}{2}(\theta_1+\theta_2+\theta_3+\theta_4) = (2n\pi + 1)\frac{\pi}{2}, \text{ n being an integer}$$

Hence $\theta_1+\theta_2+\theta_3+\theta_4 = (2n+1)\pi$, which is an odd multiple of 2p.

VALUE OF p

The main use of Gregory's series is to find the value of π to various decimal places. Some mathematicians have designed different expressions based on Gregory's series for finding the value of π. Some important cases are given below:

(a) Gregory's Series

In the Gregory's series

$$\tan^{-1} x = x - \frac{1}{3}x^3 + \frac{1}{5}x^5 - \frac{1}{7}x^7 + \ldots\infty,$$

if we put x = 1, then we have

$$\tan^{-1} 1 = 1 - \frac{1}{3} + \frac{1}{5} - \frac{1}{7} + \frac{1}{9} - \ldots\infty$$

or $$\frac{1}{4}\pi = 1 - \frac{1}{3} + \frac{1}{5} - \frac{1}{7} + \frac{1}{9} - \ldots\infty.$$

From this the value of π can be calculated. But this series does not converge rapidly and so a large number of terms will have to be taken in order to evaluate π correct to a certain decimal place. So several other series have been designed for this purpose.

(b) Euler's Series:

We have

$$\tan^{-1}\frac{1}{2} + \tan^{-1}\frac{1}{3} = \tan^{-1}\frac{\frac{1}{2}+\frac{1}{3}}{1-\frac{1}{6}} = \tan^{-1}\frac{\frac{5}{6}}{\frac{5}{6}} = \tan^{-1} 1 = \frac{\pi}{4}$$

Thus $$\frac{\pi}{4} = \tan^{-1}\frac{1}{2} + \tan^{-1}\frac{1}{3}$$

$$= \left(\frac{1}{2} - \frac{1}{3}\cdot\frac{1}{2^3} + \frac{1}{5}\cdot\frac{1}{2^5} - \ldots\right) + \left(\frac{1}{3} - \frac{1}{3}\cdot\frac{1}{5}\cdot\frac{1}{3^5} - \ldots\right),$$

expanding both $\tan^{-1}\frac{1}{2}$ and $\tan^{-1}\frac{1}{3}$ by Gregory's series because $\frac{1}{2}<1$ and $\frac{1}{3}<1$

$$=\left(\frac{1}{2}+\frac{1}{3}\right)-\frac{1}{3}\left(\frac{1}{2^3}+\frac{1}{3^3}\right)+\frac{1}{5}\left(\frac{1}{2^5}+\frac{1}{3^5}\right)-\ldots$$

From this the value of π can be calculated. This series is more rapidly convergent than the preceding one.

(c) Machin's Series:

We have

$$4\tan^{-1}\frac{1}{5}=2.2\tan^{-1}\frac{1}{5}=2\tan^{-1}\frac{\frac{2}{5}}{1-\left(\frac{1}{25}\right)}=2\tan^{-1}\frac{5}{12}$$

$$=\tan^{-1}\left\{\frac{2.\left(\frac{5}{12}\right)}{1-\left(\frac{5}{12}\right)^2}\right\}=\tan^{-1}\frac{120}{119}.$$

Now $4\tan^{-1}\frac{1}{5}-\frac{1}{4}\pi=\tan^{-1}\frac{120}{119}-\tan^{-1}1$ $\left[\because \frac{1}{4}\pi=\tan^{-1}1\right]$

$$=\tan^{-1}\frac{\frac{120}{119}-1}{1+\frac{120}{119}\times 1}=\tan^{-1}\frac{1}{239}.$$

Therefore, $\frac{\pi}{4}=4\tan^{-1}\frac{1}{5}-\tan^{-1}\frac{1}{239}$

or
$$\frac{\pi}{4}=4\left[\frac{1}{5}-\frac{1}{5}\cdot\frac{1}{5^3}+\frac{1}{5}\cdot\frac{1}{5^5}-\ldots\right]$$
$$-\left[\frac{1}{239}-\frac{1}{3}\cdot\frac{1}{(239)^3}+\frac{1}{5}\cdot\frac{1}{(239)^5}-\ldots\right],$$

expanding both the terms on the R.H.S. by Gregory's series because $\frac{1}{5}<1$ and $\frac{1}{239}<1$.

The above is Machin's series anhd it is more rapidly convergent than Euler's series.

(d) Rutherford's Series

We have

$$\tan^{-1}\frac{1}{70}-\tan^{-1}\frac{1}{99}=\tan^{-1}\frac{\frac{1}{70}-\frac{1}{99}}{1+\frac{1}{70}\cdot\frac{1}{99}}=\tan^{-1}\frac{99-70}{70.99+1}$$

$$=\tan^{-1}\frac{29}{6931}=\tan^{-1}\frac{1}{239}$$

$$=4\tan^{-1}\frac{1}{5}-\frac{\pi}{4},\text{ as shown in Machin's series.}$$

$$\therefore\ \frac{\pi}{4}=4\tan^{-1}\frac{1}{5}-\tan^{-1}\frac{1}{70}+\tan^{-1}\frac{1}{99}$$

$$i.e.,\ \frac{\pi}{4}=4\left\{\frac{1}{5}-\frac{1}{3}\frac{1}{5^3}+\frac{1}{5}\cdot\frac{1}{5^5}-\ldots\right\}-\left\{\frac{1}{70}-\frac{1}{3}\cdot\frac{1}{(70)^3}+\frac{1}{5}\cdot\frac{1}{(70)^5}-\ldots\right\}$$

$$+\left\{\frac{1}{99}-\frac{1}{3}\cdot\frac{1}{(99)^3}+\frac{1}{5}\cdot\frac{1}{(99)^5}-\ldots\right\}$$

It is Rutherford's series. This is more convenient for expansion than Machin's series and converges equally rapidly.

SOLVED PROBLEMS

Example 1:

Assuming that $\theta-n\pi=\tan\theta-\frac{1}{3}\tan^3\theta+\frac{1}{5}\tan^5\theta-\ldots$, *when* θ *lies between*

$$\left(n\pi-\frac{1}{4}\pi\right)\text{ and }\left(n\pi+\frac{1}{4}\pi\right),$$

write down the value of n when θ *lies between*

(i) $\frac{7\pi}{4}$ *and* $\frac{9\pi}{4}$

(ii) $\frac{-3\pi}{4}$ *and* $\frac{-5\pi}{4}$

(iii) $\frac{11\pi}{4}$ *and* $\frac{13\pi}{4}$

(iv) $\frac{-15\pi}{4}$ *and* $\frac{-17\pi}{4}$

(v) $\frac{-19\pi}{4}$ *and* $\frac{21\pi}{4}$.

Solution:

(i) θ lies between $\left(\frac{7\pi}{4}\right)$ and $\left(\frac{9\pi}{4}\right)$

i.e., between $\left(2\pi-\frac{1}{4}\right)$ and $\left(2\pi+\frac{1}{4}\pi\right)$. $\therefore$ n = 2

(ii) θ lies between $-\left(\frac{3\pi}{4}\right)$ and $-\left(\frac{5\pi}{4}\right)$

i.e., between $\left(-\pi+\frac{1}{4}\pi\right)$ and $\left(-\pi-\frac{1}{4}\pi\right)$. $\therefore$ n = – 1.

(iii) θ lies between $\left(\frac{111\pi}{4}\right)$ and $\left(\frac{13\pi}{4}\right)$

i.e., between $\left(3\pi-\frac{1}{4}\pi\right)$ and $\left(3\pi+\frac{1}{4}\pi\right)$, $\therefore$ n = 3.

(iv) θ lies between $-\left(\frac{15\pi}{4}\right)$ and $\left(-\frac{17\pi}{4}\right)$

i.e., between $\left(-4\pi+\frac{1}{4}\pi\right)$ and $\left(-4\pi-\frac{1}{4}\pi\right)$. $\therefore$ n = 4.

(v) θ lies between $\left(\frac{19\pi}{4}\right)$ and $\left(\frac{21\pi}{4}\right)$

i.e., between $\left(5\pi-\frac{1}{4}\pi\right)$ and $\left(5\pi+\frac{1}{4}\pi\right)$. $\therefore$ n = 5.

Example 2:

If f lies between $\left(\frac{\pi}{4}\right)$ *and* $\left(\frac{3\pi}{4}\right)$, *show that*

$$\phi=\frac{1}{2}\pi-\cot\phi+\frac{1}{3}\cot^3\phi-\frac{1}{5}\cot^5\phi+\ldots\infty.$$

Solution:

We have to prove that $\frac{1}{2}\pi-\phi=\cot\phi-\frac{1}{3}\cot^3\phi+\frac{1}{5}\cot^5\phi-\ldots\infty$.

Let $\frac{1}{2}\pi-\phi=\theta$. When $\phi=\left(\frac{\pi}{4}\right)$, we have $\theta=\frac{1}{2}\pi-\frac{1}{4}\pi=\frac{1}{4}\pi$ and when $\phi=\left(\frac{3\pi}{4}\right)$, we have $\theta=\frac{1}{2}\pi-\frac{3}{4}\pi=-\frac{1}{4}\pi$. Thus if ϕ lies between $\left(\frac{\pi}{4}\right)$ and

$\left(\frac{3\pi}{4}\right)$, then θ lies between $\left(\frac{-\pi}{4}\right)$ and $\left(\frac{\pi}{4}\right)$ *i.e.,* for θ the conditions of Gregory's series are satisfied.

Therefore we have

$$\theta = \tan\theta - \frac{1}{3}\tan^3\theta + \frac{1}{5}\tan^5\theta - \ldots\infty, \qquad \text{(by Gregory's series)}$$

$$i.e., \ \frac{1}{2}\pi - \phi = \tan\left(\frac{1}{2}\pi - \phi\right) - \frac{1}{3}\tan^3\left(\frac{1}{2}\pi - \phi\right) + \frac{1}{5}\tan^5\left(\frac{1}{2}\pi - \phi\right) - \ldots\infty.$$

$$\left[\because \theta = \frac{1}{2}\pi - \phi\right]$$

$$= \cot\phi - \frac{1}{3}\cot^3 + \frac{1}{5}\cot^5\phi - \ldots\infty.$$

This proves the result.

Example 3:

Sun the series $\frac{1}{2^3} - \frac{1}{3.2^7} + \frac{1}{5.2^{11}} - \ldots ad.inf$.

Solution:

We have

$$\frac{1}{2^3} - \frac{1}{3.2^7} + \frac{1}{5.2^{11}} + \ldots = \frac{1}{2}\left[\frac{1}{2^2} - \frac{1}{3.2^6} + \frac{1}{5.2^{10}} - \ldots\right],$$

$$\left(\text{taking } \frac{1}{2} \text{ common}\right)$$

$$= \frac{1}{2}\left[\frac{1}{(2^2)} - \frac{1}{3(2^2)^3} + \ldots\right] = \frac{1}{2}\tan^{-1}\left(\frac{1}{2^2}\right),$$

by Gregory's series because $\frac{1}{2^2} < 1$.

$$\left[\text{Note that by Gregory's series } \tan^{-1}x = x - \frac{1}{3}x^3 + \frac{1}{5}x^5 - \ldots\infty, \text{ if } |x| \le 1\right]$$

$$= \frac{1}{2}\tan^{-1}\frac{1}{4}.$$

Example 4:

Sum to infinity the series

(i) $1-\frac{1}{3.4^2}+\frac{1}{5.4^4}-...ad.inf.;$

(ii) $1-\frac{1}{3^2}+\frac{1}{5.3^2}-\frac{1}{7.3^3}+...ad.inf.$

Solution:

(i) The given series

$$1-\frac{1}{3.4^2}+\frac{1}{5.4^4}-...\text{ad.inf.}$$

$$=4\left[\frac{1}{4}-\frac{1}{3.4^3}+\frac{1}{5.4^5}-...\text{ad.inf.}\right],\qquad \text{(taking 4 common)}$$

$$=4\tan^{-1}\frac{1}{4},\text{ by Gregory's series because } \frac{1}{4}<1.$$

(ii) The given series

$$1-\frac{1}{3^2}+\frac{1}{5.3^2}-\frac{1}{7.3^3}+...=1-\frac{1}{3.3}+\frac{1}{5.3^2}-\frac{1}{7.3^3}+...$$

$$=\sqrt{3}\left[\frac{1}{\sqrt{3}}-\frac{1}{3.(\sqrt{3})^3}+\frac{1}{5(\sqrt{3})^5}-\frac{1}{7.(\sqrt{3})^7}+...\right]$$

$$=\sqrt{3}\left\{\tan^{-1}\left(\frac{1}{\sqrt{3}}\right)\right\},\text{ by Gregory's series because } \frac{1}{\sqrt{3}}<1$$

$$=\sqrt{3}.\left(\frac{\pi}{6}\right)=\frac{\left(\frac{\pi}{\sqrt{3}}\right)}{6}.$$

Note: This question can also be written as

Prove that $\pi=2\sqrt{3}\left\{1-\frac{1}{3^2}+\frac{1}{5.3^2}-\frac{1}{7.3^3}+...\right\}$

Example 5:

Prove that

$$\frac{\pi}{4}=\frac{17}{21}-\frac{713}{81\times343}+...+\frac{(-1)^{n+1}}{2n-1}\left\{\frac{2}{3}9^{1-n}+7^{1-2n}\right\}+...$$

Solution:

The nth term of the given series on the R.H.S. is

$$T_n = \frac{(-1)^{n+1}}{2n-1}\left\{\frac{2}{3}.\left(3^2\right)^{1-n} + 7^{1-2n}\right\}$$

$$= \frac{(-1)^{n+1}}{2n-1}\left\{\frac{2}{3}.3^{2-2n} + 7^{1-2n}\right\} = \frac{(-1)^{n+1}}{(2n-1)}\left[\frac{2}{3^{2n-1}} + \frac{1}{7^{2n-1}}\right].$$

Putting n = 1, 2, 3, ..., etc., we have

first term = $T_1 = \left(\frac{2}{3} + \frac{1}{7}\right);$

second term = $T_2 = -\frac{1}{3}\left(\frac{2}{3^3} + \frac{1}{7^3}\right) = -\frac{2}{3.3^3} - \frac{1}{3.7^3};$

third term = $T_3 = \frac{1}{5}\left(\frac{2}{3^5} + \frac{1}{7^5}\right) = \frac{2}{5}.\frac{1}{3^5} + \frac{1}{5.7^5};$

and so on.

∴ the sum of the series on the R.H.S.

$= T_1 + T_2 + T_3 + \ldots$ ad.inf.

$$= \left(\frac{2}{3} + \frac{1}{7}\right) - \left(\frac{2}{3.3^3} + \frac{1}{3.7^3}\right) + \left(\frac{2}{5.3^5} + \frac{1}{5.7^5}\right) - \ldots$$

$$= 2\left(\frac{1}{3} - \frac{1}{3.3^3} + \frac{1}{5.3^5} - \ldots\right) + \left(\frac{1}{7} - \frac{1}{3.7^3} + \frac{1}{5.7^5} - \ldots\right)$$

$$= 2\tan^{-1}\left(\frac{1}{3}\right) + \tan^{-1}\left(\frac{1}{7}\right), \text{ (by Gregory's series)}$$

$$= \tan^{-1}\left\{\frac{\frac{1}{3} + \frac{1}{3}}{1 - \frac{1}{3}.\frac{1}{3}}\right\} + \tan^{-1}\frac{1}{7} = \tan^{-1}\frac{3}{4} + \tan^{-1}\frac{1}{7}$$

$$= \tan^{-1}\left\{\frac{\frac{3}{4} + \frac{1}{7}}{1 - \frac{3}{4}.\frac{1}{7}}\right\} = \tan^{-1}1 = \frac{\pi}{4}.$$

Example 6:

Express tan⁻¹ (cos θ + i sin θ) in the form A + iB and deduce that

(i) $\cos\theta - \frac{1}{3}\cos 3\theta + \frac{1}{5}\cos 5\theta - \ldots = \pm\frac{1}{4}\pi;$ *and*

(ii) $\sin\theta - \frac{1}{3}\sin 3\theta + \frac{1}{5}\sin 5\theta - ... = \frac{1}{2}\log\left\{\pm\tan\left(\frac{1}{4}\pi + \frac{1}{2}\theta\right)\right\}.$

Solution:

Let $\tan^{-1}(\cos\theta + i\sin\theta) = A + iB$

so that $\tan^{-1}(\cos\theta - i\sin\theta) = A - iB$, (complex conjugates).

$\therefore 2A = (A + iB) + (A - iB)$

$= \tan^{-1}(\cos\theta + i\sin\theta) + \tan^{-1}(\cos\theta - i\sin\theta)$

$$= \tan^{-1}\left\{\frac{(\cos\theta + i\sin\theta) + (\cos\theta - i\sin\theta)}{1 - (\cos\theta + i\sin\theta)(\cos\theta - i\sin\theta)}\right\}$$

$$= \tan^{-1}\left\{\frac{2\cos\theta}{1 - (\cos^2\theta + \sin^2\theta)}\right\} = \tan^{-1}\left(\frac{2\cos\theta}{1-1}\right) = \tan^{-1}\left(\frac{2\cos\theta}{0}\right)$$

$$= \tan^{-1}(\pm\infty) = \pm\frac{1}{2}\pi. \qquad [\because \theta \text{ lies between } -\pi \text{ and } \pi]$$

Therefore, $A = \pm\frac{1}{4}\pi.$ ...(1)

Again $2iB = (A + iB) - (A - iB) = \tan^{-1}(\cos\theta + i\sin\theta) - \tan^{-1}(\cos\theta - i\sin\theta)$

$$= \tan^{-1}\left\{\frac{(\cos\theta + i\sin\theta) - (\cos\theta - i\sin\theta)}{1 + (\cos\theta + i\sin\theta)(\cos\theta - i\sin\theta)}\right\}$$

$$= \tan^{-1}\left\{\frac{2i\sin\theta}{1 + (\cos^2\theta + \sin^2\theta)}\right\}$$

$$= \tan^{-1}\left\{\frac{2i\sin\theta}{1+1}\right\} = \tan^{-1}(i\sin\theta)$$

or $\tan(2iB) = i\sin\theta$ or $\tanh 2B = \sin\theta$

or $\frac{e^{2B} - e^{-2B}}{e^{2B} + e^{-2B}} = \frac{\sin\theta}{1}$ or $\frac{2e^{2B}}{2e^{-2B}} = \frac{1+\sin\theta}{1-\sin\theta}$

[By componendo and dividendo]

i.e., $$e^{4B} = \frac{1+\sin\theta}{1-\sin\theta} = \frac{\left\{\cos\left(\frac{\theta}{2}\right) + \sin\left(\frac{\theta}{2}\right)\right\}^2}{\left\{\cos\left(\frac{\theta}{2}\right) - \sin\left(\frac{\theta}{2}\right)\right\}} = \pm\tan\left(\frac{\pi}{4} + \frac{\theta}{2}\right).$$

$$\therefore \qquad B = \frac{1}{2}\log\left\{\pm\tan\left(\frac{\pi}{4}+\frac{\theta}{2}\right)\right\} \qquad ...(2)$$

Now $\tan^{-1}(\cos\theta + i\sin\theta) = A + iB$

or $\qquad \tan^{-1}(e^{i\theta}) = A + iB$

or $\qquad e^{i\theta} - \frac{1}{3}e^{3i\theta} + \frac{1}{5}e^{5i\theta} - ... = A + iB$, using Gregory's series

or $(\cos\theta + i\sin\theta) - \frac{1}{3}(\cos 3\theta + i\sin 3\theta) + \frac{1}{5}(\cos 5\theta + i\sin 5\theta) - ... = A + iB$

or $\left(\cos\theta - \frac{1}{3}\cos 3\theta + \frac{1}{5}\cos 5\theta - ...\right) + i\left(\sin\theta - \frac{1}{3}\sin 3\theta + \frac{1}{5}\sin 5\theta - ...\right)$

$= A + iB = \pm\frac{1}{4}\pi + \frac{1}{2}i\log\left\{\pm\tan\left(\frac{1}{4}\pi + \frac{1}{2}\theta\right)\right\}$, by (1) and (2)

Equating real and and imaginary parts, we have

$$\cos\theta - \frac{1}{3}\cos 3\theta + \frac{1}{5}\cos 5\theta - ... = \pm\frac{1}{4}$$

and $\sin\theta - \frac{1}{3}\sin 3\theta + \frac{1}{5}\sin 5\theta - ... = \frac{1}{2}\log\left\{\pm\tan\left(\frac{1}{4}\pi + \frac{1}{2}\theta\right)\right\}$.

Example 8:

Expand $\dfrac{1 + a\cos\theta}{1 + 2a\cos\theta + a^2}$ *in a series of cosines of multiples of* θ, *where* $a < 1$.

Solution:

We have

$$\frac{1 + a\cos\theta}{1 + 2a\cos\theta + a^2} = \frac{1 + \frac{1}{2}a\left(e^{i\theta} + e^{-i\theta}\right)}{1 + a\left(e^{i\theta} + e^{-i\theta}\right) + a^2} \qquad \left[\because \cos\theta = \frac{1}{2}\left(e^{i\theta} + e^{-i\theta}\right)\right]$$

$$= \frac{1}{2}\cdot\frac{2 + a\left(e^{i\theta} + e^{-i\theta}\right)}{1 + a\left(e^{i\theta} + e^{-i\theta}\right) + a^2\, e^{i\theta}.e^{-i\theta}}$$

$$= \frac{1}{2}\frac{\left(1 + ae^{i\theta}\right) + \left(1 + ae^{-i\theta}\right)}{\left(1 + ae^{i\theta}\right)\left(1 + ae^{-i\theta}\right)} = \frac{1}{2}\left[\frac{1}{\left(1 + ae^{i\theta}\right)} + \frac{1}{\left(1 + ae^{-i\theta}\right)}\right]$$

$$=\frac{1}{2}\left[\left(1+a e^{i\theta}\right)^{-1}+\left(1+a e^{-i\theta}\right)^{-1}\right]$$

$$=\frac{1}{2}[(1 - a\ e^{i\theta} + a^2\ a^{2i\theta} - a^3\ e^{3i\theta} + \ldots)$$

$$+ (1 - a\ e^{-i\theta} + a^2\ e^{-2i\theta} - a^3\ e^{-3i\theta} + \ldots)],$$

expanding by binomial theorem which is justified since $a < 1$

$$=11-a\left(\frac{e^{i\theta}+e^{-i\theta}}{2}\right)+a^2\left(\frac{e^{2i\theta}+e^{-2i\theta}}{2}\right)-a^3\left(\frac{e^{3i\theta}+e^{-3i\theta}}{2}\right)+\ldots$$

$$= 1 - a\cos\theta + a^2\cos 2\theta - a^3\cos 3\theta + \ldots$$

Example 8:

Expand $\dfrac{\cos\theta - a\cos(\theta-\phi)}{1-2a\cos\phi + a^2}$ *in an infinite series if* $a < 1$.

Solution:

We have

$$\frac{\cos\theta - a\cos(\theta-\phi)}{1-2a\cos\phi+a^2}=\frac{1}{2}\cdot\frac{\left(e^{i\theta}+e^{-i\theta}\right)-a\left\{e^{i(\theta-\phi)}+e^{-i(\theta-\phi)}\right\}}{1-a\left(e^{i\phi}+e^{-i\phi}\right)+a^2}$$

$$=\frac{1}{2}\cdot\frac{e^{i\theta}\left(1-a e^{-i\phi}\right)+e^{-i\theta}\left(1-a e^{i\phi}\right)}{\left(1-a e^{-i\phi}\right)\left(1-a e^{i\phi}\right)}$$

$$=\frac{1}{2}\cdot\left[e^{i\theta}\left(1-a e^{i\phi}\right)^{-1}+e^{-i\theta}\left(1-a e^{-i\phi}\right)^{-1}\right]$$

$$=\frac{1}{2}[e^{i\theta}\ (1 + a\ e^{i\phi} + a^2\ e^{2i\phi} + a^3\ e^{3i\phi} + \ldots)$$

$$+ e^{-i\theta}\ (1 + a\ e^{i\phi} + a^2\ e^{-2i\phi} + a^3\ e^{-3i\phi} + \ldots)].$$

on expanding by binomial theorem which is valid because $a < 1$

$$=\frac{1}{2}\left[e^{i\theta}+e^{-i\theta}\right]+\frac{1}{2}a\left[e^{i(\theta+\phi)}+e^{-i(\theta+\phi)}\right]+\frac{1}{2}a^2\left[e^{i(\theta+2\phi)}+e^{-i(\theta+2\phi)}\right]+\ldots\infty$$

$$= \cos\theta + a\cos(\theta+\phi) + a^2\cos(\theta+2\phi) + a^3\cos(\theta+3\phi) + \ldots\ \infty.$$

Example 9:

Expand $e^{ax}\sin bx$ *in a series of ascending powers of* x. *Also find the coefficient of* x^n *in this expansion.*

Solution:

We have $e^{ax} \sin bx$

$$= \frac{1}{2i} e^{ax}\left[e^{ibx} - e^{-ibx}\right] = \frac{1}{2i}\left[e^{(a+ib)x} - e^{(a-ib)x}\right]$$

$$= \frac{1}{2i}\left[\left\{1+(a+ib)x+\frac{(a+ib)^2 x^2}{2!}+\ldots\right\}-\left\{a+(a-ib)x+\frac{(a-ib)^2 x^2}{2!}+\ldots\right\}\right]$$

$[\because e^x = 1 + x/1! + x^2/2! + x^3/3! + \ldots]$

$$= \frac{1}{2i}\left[x\{(a+ib)-(a-ib)\}+\frac{x^2}{2!}\left\{(a+ib)^2-(a-ib)^2\right\}+\ldots\right] \qquad \ldots(1)$$

Now put $a = r \cos \theta$ and $b = r \sin \theta$, so that

$r = \sqrt{(a^2+b^2)}$ and $\theta = \tan^{-1}\left(\frac{b}{a}\right)$.

Then the coefficient of x^n in (1)

$$= \frac{1}{2i}\frac{1}{n!}\left[(a+ib)^n-(a-ib)^n\right] = \frac{1}{2i(n!)}$$

$$[(r \cos \theta + ir \sin \theta)^n - (r \cos \theta - ir \sin \theta)^n]$$

$$= \frac{r^n}{2i(n!)}\left[(\cos n\theta + i \sin n\theta)-(\cos n\theta - i \sin n\theta)\right]$$

$$= \frac{r^n}{2i(n!)} 2i \sin n\theta = \frac{r^n}{n!}\sin n\theta.$$

$\therefore$ from (1), we have

$$e^{ax} \sin bx = x (r \sin \theta) + \frac{x^2 r^2 \sin 2\theta}{2!}+\frac{x^3 r^3 \sin 3\theta}{3!}+\ldots+\frac{x^n r^n \sin n\theta}{n!}+\ldots$$

Also the coefficient of x^n

$$= \frac{r^n}{n!}\sin n\theta = \frac{(a^2+b^2)^{n/2}}{n!}\sin\left\{n \tan^{-1}\frac{b}{a}\right\}.$$

The expansion is valid for all values of a, b, x.

Example 10:

Expand $e^{a \cos \phi} \cos(\theta + a \sin \phi)$ in an infinite series.

Solution:

We have

$e^{a\cos\phi}\cos(\theta + a\sin\phi)$

$$= e^{a\cos\phi}.\left[\frac{e^{i(\theta+a\sin\phi)} + e^{-i(\theta+a\sin\phi)}}{2}\right] \qquad \textbf{(Note)}$$

$$= \frac{1}{2}\left\{e^{i\theta}.e^{a(\cos\phi+i\sin\phi)} + e^{-i\theta}.e^{a(\cos\phi-i\sin\phi)}\right\}$$

$$= \frac{1}{2}\left\{e^{i\theta}.e^{ae^{i\phi}} + e^{-i\theta}.e^{ae^{-i\phi}}\right\} \qquad (\because \cos\phi + i\sin\phi = e^{i\phi})$$

$$= \frac{1}{2}\left[e^{i\theta}\left(1 + ae^{i\phi} + \frac{1}{2!}a^2e^{2i\phi} + \frac{1}{3!}a^3e^{3i\phi} + \ldots\right)\right.$$

$$\left. + e^{-i\theta}\left(1 + ae^{-i\phi} + \frac{1}{2!}^2 e^{-2i\phi} + \frac{1}{3!}a^3e^{-3i\phi} + \ldots\right)\right]$$

$$= \frac{1}{2}\left[\left\{e^{i\theta} + e^{-i\theta}\right\} + a\left\{e^{i(\theta+\phi)}\right\} + e^{-i(\theta+\phi)}\right.$$

$$\left. + \frac{1}{2!}a^2\left\{e^{i(\theta+2\phi)} + e^{-i(\theta+2\phi)}\right\} + \frac{1}{3!}a^3\left\{e^{i(\theta+3\phi)} + e^{-i(\theta+3\phi)} + \ldots\right\}\right]$$

$= \cos\theta + a\cos(\theta + \phi) + (1/2!)\,a^2\cos(\theta + 2\phi) + (1/3!)\,a^3\cos(\theta + 3\phi) + \ldots$

Example 11:

Expand log cos $\left(\frac{1}{4}\pi + \theta\right)$ *in a series of sine and cosine of multiples of* θ.

Solution:

We have

$$\log\cos\left(\frac{1}{4}\pi + \theta\right) = \frac{1}{2}\log\cos^2\left(\frac{1}{4}\pi + \theta\right)$$

$$= \frac{1}{2}\log\frac{2\cos^2\left(\frac{1}{4}\pi + \theta\right)}{2} = \frac{1}{2}\log\frac{1 + \cos\left(\frac{1}{2}\pi + 2\theta\right)}{2}$$

$$= \frac{1}{2}\log\frac{1 - \sin 2\theta}{2} = \frac{1}{2}\log\left\{\frac{1 - \left(\frac{1}{2i}\right)\left(e^{2i\theta} - e^{-2i\theta}\right)}{2}\right\}$$

$$= \frac{1}{2}\log\left\{\frac{2-\left(\frac{i}{i^2}\right)\left(e^{2i\theta}-e^{-2i\theta}\right)}{4}\right\} = \frac{1}{2}\log\frac{2+i\left(e^{2i\theta}-e^{-2i\theta}\right)}{4}$$

$$= \frac{1}{2}\log\frac{\left(1+ie^{2i\theta}\right)\left(1-ie^{-2i\theta}\right)}{2^2} \qquad \textbf{(Note)}$$

$$= \frac{1}{2}\log\left\{\left(1+ie^{2i\theta}\right)\left(1+ie^{2i\theta}\right)\left(1-ie^{-2i\theta}\right)\right\} - \frac{1}{2}.2\log 2$$

$$= \frac{1}{2}\left[\log\left(1+ie^{2i\theta}\right)+\log\left(1-ie^{-2i\theta}\right)\right]-\log 2$$

$$= \frac{1}{2}\left[\left(ie^{2i\theta}-\frac{1}{2}.i^2e^{4i\theta}+\frac{1}{3}.i^3e^{6i\theta}-\ldots\right)\right.$$

$$\left.-\left(ie^{-2i\theta}+\frac{1}{2}i^2e^{-4i\theta}+\frac{1}{3}i^3e^{-6i\theta}+\ldots\right)\right]-\log 2,$$

on expanding by logarithmic expansion

$$= \left[i\cdot\left(\frac{e^{2i\theta}-e^{-2i\theta}}{2}\right)+\frac{1}{2}\cdot\left(\frac{e^{4i\theta}+e^{-4i\theta}}{2}\right)-\frac{1}{3}i\cdot\left(\frac{e^{6i\theta}-e^{-6i\theta}}{2}\right)+\ldots\right]-\log 2$$

$$= \left[i.i\sin 2\theta+\frac{1}{2}.\cos 4\theta-\frac{1}{3}i.i\sin 6\theta+\ldots\right]-\log 2$$

$$= -\sin 2\theta+\frac{1}{2}\cos 4\theta+\frac{1}{3}\sin 6\theta+\ldots-\log 2$$

$$= -\log 2-\sin 2\theta+\frac{1}{2}\cos 4\theta+\frac{1}{3}\sin 6\theta-\frac{1}{4}\cos 8\theta\ldots\text{ad.inf.}$$

Example 12:

Prove that

$$\log\sin\theta = -\log 2-\cos 2\theta-\frac{1}{2}\cos 4\theta-\frac{1}{3}\cos 6\theta-\ldots\ \text{ad.inf.}$$

Hence prove that if a and b the sides of a plane triangle, A and B the opposite angles,

$$\log b-\log a = (\cos 2A-\cos 2B)+\frac{1}{2}\cos 4\theta-\frac{1}{3}\ \text{co}\ 6\theta-\ldots\ \text{ad.inf.}$$

Solution:

We have

$$\log \sin\theta = \frac{1}{2}\log\sin^2\theta = \frac{1}{2}\log\left\{\frac{(1-\cos 2\theta)}{2}\right\} = \frac{1}{2}\log\left\{\frac{1-\frac{1}{2}\left(e^{2i\theta}+e^{-2i\theta}\right)}{2}\right\}$$

$$= \frac{1}{2}\log\left\{\frac{2-\left(e^{2i\theta}+e^{-2i\theta}\right)}{4}\right\} = \frac{1}{2}\log\left\{\frac{\left(1-e^{2i\theta}\right)\left(1-e^{-2i\theta}\right)}{4}\right\}$$

$$= \frac{1}{2}\log\left\{\left(1-e^{2i\theta}\right)\left(1-e^{-2i\theta}\right)\right\} - \frac{1}{2}\log 4$$

$$= \frac{1}{2}\left[\log\left(1-e^{2i\theta}\right)+\log\left(1-e^{-2i\theta}\right)\right] - \log 2$$

$$= \frac{1}{2}\left[\left(-e^{2i\theta}-\frac{1}{2}e^{4i\theta}-\ldots\right)+\left(-e^{-2i\theta}-\frac{1}{2}e^{-4i\theta}\right)-\ldots\right] - \log 2$$

$$= -\frac{1}{2}\left[\left(e^{2i\theta}+e^{-2i\theta}\right)+\frac{1}{2}\left(e^{4i\theta}+e^{-4i\theta}\right)+\ldots\right] - \log 2$$

$$= \left[-\cos 2\theta - \frac{1}{2}\cos 4\theta - \frac{1}{3}\cos 6\theta - \ldots\right] - \log 2$$

$$= -\log 2 - \cos 2\theta - \frac{1}{2}\cos 4\theta - \frac{1}{3}\cos 6\theta - \ldots \text{ad.inf.} \qquad \ldots(1)$$

This proves the first result.

Also we know that

$$\frac{\sin A}{a} = \frac{\sin B}{b} = k(\text{say}).$$

$\therefore$ ak = sin A, and bk = sin B.

Using (1), we have

$$\log ak = \log \sin A = -\log 2 - \cos 2A - \frac{1}{2}\cos 4A - \ldots \qquad \ldots(2)$$

$$\text{and } \log bk = \log \sin B = -\log 2 - \cos 2B - \frac{1}{2}\cos 4B - \ldots$$

Subtracting (2) and (3), we have

$$\log bk - \log ak = (\cos 2A - \cos 2B) + \frac{1}{2}(\cos 4A - \text{co } 4B) + \ldots$$

$$\text{or } (\log b + \log k) - (\log a + \log k) = (\cos 2A - \cos 2B) + \frac{1}{2}(\cos 4A - \cos 4B) + \ldots$$

or log b – log a = (cos 2A – cos 2B) + $\frac{1}{2}$ (cos 4A – cos 4B) + ...

This proves the second result.

Example 13:

In any triangle where a > b, prove that

$$\log c = \log a - \frac{b}{a}\cos C - \frac{1}{2}\frac{b^2}{a^2}\cos 2C - \frac{1}{3}\frac{b^3}{a^3}\cos 3C - \ldots$$

Solution:

We know that

$$c^2 = a^2 + b^2 - 2ab\cos C = a^2\left[1 + \frac{b^2}{a^2} - 2\frac{b}{a}\cos C\right]$$

$$= a^2\,[1 + x^2 - 2x\cos C], \text{ putting } \left(\frac{b}{a}\right) = x$$

$$= a^2\left[1 + x^2 - 2x\cdot\frac{1}{2}\left(e^{iC} + e^{-iC}\right)\right] = a^2\left\{\left(1 - xe^{iC}\right)\left(1 - xe^{-iC}\right)\right\}.$$

Taking logarithm of both sides, we have

$$2\log c = 2\log a + \log(1 - xe^{iC}) + \log(1 - xe^{-iC})$$

$$= 2\log a + \left(-xe^{iC} - \frac{1}{2}x^2e^{2iC} - \frac{1}{3}x^3e^{3iC} - \ldots\right)$$

$$+\left(-xe^{-iC} - \frac{1}{2}x^2e^{-2iC} - \frac{1}{3}x^3e^{-3iC} - \ldots\right),$$

expanding by logarithmic expansion since x < 1, a being given to be greater than b

$$= 2\log a - x\left(e^{iC} + e^{-iC}\right) - \frac{1}{2}x^2\left(e^{2iC} + e^{-2iC}\right) - \frac{1}{3}x^3\left(e^{3iC} - e^{-3iC}\right) - \ldots$$

$$= 2\log a - 2x\cos C - \frac{1}{2}x^2.2\cos 2C - \frac{1}{3}x^3.2\cos 3C - \ldots$$

$$= \log c - \log a - \frac{b}{a}\cos C - \frac{1}{2}\frac{b^2}{a^2}\cos 2C - \frac{1}{3}\frac{b^3}{a^3}\cos 3C - \ldots$$

[Putting x = b/a]

Example 14:

If $\tan\theta = \dfrac{a\sin\alpha}{1 + a\cos\alpha}$, *prove that* $\theta = a\sin\alpha - \frac{1}{2}a^2\sin 2\alpha + \frac{1}{3}a^3\sin 3\alpha - \ldots$

Solution:

We have $\tan\theta = \dfrac{a\sin\alpha}{1+a\cos\alpha}$

$$\therefore \quad \frac{\sin\theta}{\cos\theta} = \frac{a\sin\alpha}{1+a\cos\alpha}$$

or $\dfrac{i\sin\theta}{\cos\theta} = \dfrac{ia\sin\alpha}{1+a\cos\alpha}$, multiplying both sides by i.

Applying componendo and dividendo, we have

$$\frac{\cos\theta + i\sin\theta}{\cos\theta - i\sin\theta} = \frac{1+a\cos\alpha + ia\sin\alpha}{1+a\cos\alpha - ia\sin\alpha}$$

or $\dfrac{e^{i\theta}}{e^{-i\theta}} = \dfrac{1+ae^{i\alpha}}{1+ae^{-i\alpha}}$ or $e^{2i\theta} = \dfrac{1+ae^{i\alpha}}{1+ae^{-i\alpha}}$.

Taking logarithm of both sides, we have

$2i\theta = \log(1 + ae^{i\alpha}) - \log(1 + ae^{-i\alpha})$

$$= \left(ae^{i\alpha} - \frac{1}{2}a^2e^{i2\alpha} + \frac{1}{3}a^3e^{i3\alpha} - \ldots\right) - \left(ae^{-i\alpha} - \frac{1}{2}a^2e^{-2\alpha} + \frac{1}{3}a^3e^{-3i\alpha} - \ldots\right),$$

the logarithmic expansions are valid if $|a| < 1$

$$= a\left(e^{i\alpha} - e^{-i\alpha}\right) - \frac{1}{2}a^2\left(e^{i2\alpha} - e^{-2\alpha}\right) + \frac{1}{3}a^3\left(e^{i3\alpha} - e^{-i3\alpha}\right) - \ldots$$

$$\therefore \quad \theta = a\left(\frac{e^{i\alpha} - e^{-i\alpha}}{2i}\right) - \frac{1}{2}a^2\frac{e^{i2\alpha} - e^{-i2\alpha}}{2i} + \frac{1}{3}a^3\frac{e^{i3\alpha} - e^{-i3\alpha}}{2i} - \ldots$$

$$= a\sin\alpha - \frac{1}{2}a^2\sin 2\alpha + \frac{1}{3}a^3\sin 3\alpha - \ldots$$

Example 15:

If $\tan\theta = x + \tan\alpha$, *show that*

$$\theta = \alpha + x\cos^2\alpha - \frac{1}{2}x^2\cos^2\alpha\sin 2\alpha - \frac{1}{3}x^3\cos^3\alpha\cos 3\alpha + \ldots$$

Solution:

We have $\tan\theta = x + \tan\alpha$

i.e., $\dfrac{1}{i}\cdot\dfrac{e^{i\theta} - e^{-i\theta}}{e^{i\theta} + e^{-i\theta}} = x + \dfrac{\sin\alpha}{\cos} = \dfrac{x\cos\alpha + \sin\alpha}{\cos\alpha}$

or $\dfrac{e^{i\theta} - e^{-i\theta}}{e^{i\theta} - e^{-i\theta}} = \dfrac{ix\cos\alpha + i\sin\alpha}{\cos\alpha}$

Applying componendo and dividendo, we have

$$\frac{2e^{i\theta}}{2e^{-i\theta}}=\frac{\cos\alpha+(ix\cos\alpha+i\sin\alpha)}{\cos\alpha-(ix\cos\alpha+i\sin\alpha)}=\frac{ix\cos\alpha+(\cos\alpha+i\sin\alpha)}{(\cos\alpha-i\sin\alpha)-ix\cos\alpha}$$

or $$e^{2i\theta}=\frac{ix\cos\alpha+e^{i\alpha}}{e^{-i\alpha}-ix\cos\alpha}=\frac{e^{i\alpha}\left(1+ixe^{-i\alpha}\cos\alpha\right)}{e^{-i\alpha}\left(1-ixe^{i\alpha}\cos\alpha\right)}.$$

Taking logarithm of both sides, we get

$2i\theta = 2i\alpha + \log(1 + ix\, e^{-i\alpha}\cos\alpha) - \log(1 - ix\, e^{i\alpha}\cos\alpha)$

$$=2i\alpha+\left[ixe^{-i\alpha}\cos\alpha-\frac{i^2x^2e^{-2i\alpha}\cos^2\alpha}{2}+\frac{i^3x^3e^{-3i\alpha}\cos^3\alpha}{3}-...\right]$$

$$-\left[-ixe^{i\alpha}\cos\alpha-\frac{i^2x^2e^{2i\alpha}\cos^2\alpha}{2}-\frac{i^3x^3e^{3i\alpha}\cos^3\alpha}{2}-...\right]$$

$$=2i\alpha+ix\cos\alpha.\left(e^{i\alpha}+e^{-i\alpha}\right)+\frac{1}{2}i^2x^2\cos^2\alpha\left(e^{2i\alpha}-e^{-2i\alpha}\right)$$

$$+\frac{1}{3}i^3x^3\cos^3\alpha\left(e^{3i\alpha}-e^{-3i\alpha}\right)+...$$

$$= 2i\alpha + ix\cos\alpha\,(2\cos\alpha) - \frac{1}{2}x^2\cos^2\alpha\,(2i\sin 2\alpha) - \frac{1}{3}ix^3$$
$$\cos^3\alpha\,(2\cos 3\alpha) + ...$$

or $$2i\theta = 2i\left[\alpha + x\cos^2\alpha - \frac{1}{2}x^2\cos^2\alpha\sin 2\alpha - \frac{1}{3}x^3\cos^3\alpha\right.$$
$$\left.\cos 3\alpha + ...\right]$$

or $$\theta = \alpha + \cos^2\alpha - \frac{1}{2}x^2\cos^2\sin 2\alpha - \frac{1}{3}x^3\cos^3\alpha\cos 3\alpha + ...\infty.$$

Example 16:

If $\cot\phi = (1/x) + \cot\theta$, *prove that*

$$\phi=\frac{x}{\sin\theta}\sin\theta-\frac{1}{2}\frac{x^2}{\sin^2\theta}.\sin 2\theta+\frac{1}{3}\frac{x^3}{\sin^3\theta}\sin 3\theta-...$$

Solution:

We have

$$\cot\phi=\frac{1}{x}+\cot\theta \text{ or } \frac{\cos\phi}{\sin\phi}=\frac{1}{x}+\frac{\cos\theta}{\sin\theta}$$

or $$\frac{\left(e^{i\phi}+e^{-i\phi}\right)/2}{\left(e^{i\phi}-e^{-i\phi}\right)/2i}=\frac{\sin\theta+x\cos\theta}{x\sin\theta}.$$

or $\dfrac{e^{i\phi}+e^{-i\phi}}{e^{i\phi}-e^{-i\phi}}=\dfrac{\sin\theta+x\cos\theta}{ix\sin\theta}$.

Applying componedo and dividendo, we get

$$\frac{\left(e^{i\phi}+e^{-i\phi}\right)+\left(e^{i\phi}-e^{-i\phi}\right)}{\left(e^{i\phi}+e^{-i\phi}\right)-\left(e^{i\phi}-e^{-i\phi}\right)}=\frac{(\sin\theta+x\cos\theta)+ix\sin\theta}{(\sin\theta+x\cos\theta)-ix\sin\theta}$$

or $\dfrac{2e^{i\phi}}{2e^{-i\phi}}=\dfrac{\sin\theta+x(\cos\theta+i\sin\theta)}{\sin\theta+x(\cos\theta-i\sin\theta)}=\dfrac{\sin\theta+xe^{i\theta}}{\sin\theta+xe^{-i\theta}}$

or $e^{2i\phi}=\dfrac{\sin\theta\left[1+\left(\dfrac{x}{\sin\theta}\right)e^{i\theta}\right]}{\sin\theta\left[1+\left(\dfrac{x}{\sin\theta}\right)e^{-i\theta}\right]}=\dfrac{1+Ae^{i\theta}}{1+Ae^{-i\theta}}$, where $A=\dfrac{x}{\sin\theta}$.

Taking logarithm of both sides, we have

$$2i\phi=\log(1+Ae^{i\theta})-\log(1+Ae^{-i\theta})$$

$$=\left(Ae^{i\theta}-\frac{1}{2}A^2e^{2i\theta}+\frac{1}{3}A^3e^{3i\theta}-\ldots\right)$$

$$-\left(Ae^{-i\theta}-\frac{1}{2}A^2e^{-2i\theta}+\frac{1}{3}A^3e^{-3i\theta}-\ldots\right)$$

$$=A\left(e^{i\theta}-e^{-i\theta}\right)-\frac{1}{2}A^2\left(e^{2i\theta}-e^{-2i\theta}\right)+\frac{1}{3}A^3\left(e^{3i\theta}-e^{-3i\theta}\right)-\ldots$$

$$=A(2i\sin\theta)-\frac{1}{2}A^2(2i\sin2\theta)+\frac{1}{3}A^3(2i\sin3\theta)-\ldots$$

or $\phi=A\sin\theta-\dfrac{1}{2}A^2\sin2\theta+\dfrac{1}{3}A^3\sin3\theta-\ldots$

$$\phi=\frac{x}{\sin\theta}\cdot\sin\theta-\frac{1}{2}\frac{x^2}{\sin^2\theta}\sin2\theta+\frac{1}{3}\frac{x^3}{\sin^3\theta}\sin3\theta-\ldots$$

[Putting $A=(x/\sin\theta)$]

EXPANSIONS

4. *Expansion of* $\dfrac{x\sin\theta}{1-2x\cos\theta+x^2}$ *in a series of multiples of* θ *where* $x<1$.

or Expansion of $\dfrac{x\sin\theta}{1-2x\cos\theta+x^2}$ *in a series of ascending powers of x.*

We know that

$\sin\theta = (e^{i\theta} - e^{-i\theta})/2i$ and $\cos\theta = (e^{i\theta} + e^{-i\theta})/2$.

$$\therefore \quad \frac{x\sin\theta}{1-2x\cos\theta+x^2} = \frac{x\{(e^{i\theta}-e^{-i\theta})/2i\}}{1-2x\cdot\frac{1}{2}(e^{i\theta}+e^{-i\theta})+x^2}$$

$$= \frac{1}{2i}\cdot\frac{x(e^{i\theta}-e^{-i\theta})}{1-x(e^{i\theta}+e^{-i\theta})+x^2} = \frac{1}{2i}\cdot\frac{x(e^{i\theta}-e^{-i\theta})}{(1-xe^{i\theta})(1-xe^{-i\theta})} \qquad \textbf{(Note)}$$

$$= \frac{1}{2i}\left[\frac{1}{(1-xe^{i\theta})} - \frac{1}{(1-xe^{-i\theta})}\right], \qquad \text{(by partial fractions)}$$

$$= \left(\frac{1}{2i}\right)[(1 - x\,e^{i\theta})^{-1} - (1 - x\,e^{-i\theta})^{-1}]$$

$$= \left(\frac{1}{2i}\right)[(1 + x\,e^{i\theta} + x^2\,e^{2i\theta} + x^3\,e^{3i\theta} + \ldots)$$
$$- (1 + xe^{-i\theta} + x^2\,e^{-2i\theta} + x^3\,e^{-3i\theta} + \ldots)],$$

[expanding both $(1 - xe^{i\theta})^{-1}$ and $(1 - xe^{-i\theta})^{-1}$ by binomial theorem which is justified because $|xe^{i\theta}| = |x|\,|e^{i\theta}| = |x| \cdot |\cos\theta + i\sin\theta| = |x| < 1$, x being given to be < 1, and similarly $|xe^{-i\theta}| < 1$]

$$= \frac{1}{2i}\left[x(e^{i\theta}-e^{-i\theta})+x^2(e^{2i\theta}-e^{-2i\theta})+x^3(e^{3i\theta}-e^{-3i\theta})+\ldots\right]$$

$$= x\left(\frac{e^{i\theta}-e^{-i\theta}}{2i}\right)+x^2\left(\frac{e^{2i\theta}-e^{-2i\theta}}{2i}\right)+x^3\left(\frac{e^{3i\theta}-e^{-3i\theta}}{2i}\right)+\ldots$$

$= x\sin\theta + x^2\sin 2\theta + x^3\sin 3\theta + \ldots$ ad.inf.

5. *If $x < 1$, prove that*

$$\frac{x\cos\theta}{1-2x\sin\theta+x^2} = x\cos\theta + x^2\sin 2\theta - x^3\cos 3\theta - x^4\sin 4\theta + \ldots \textit{ ad.inf.}$$

We have

$$\frac{x\cos\theta}{1-2x\sin\theta+x^2} = \frac{x.\frac{1}{2}(e^{i\theta}+e^{-i\theta})}{1-2x\{(e^{i\theta}-e^{-i\theta})/2i\}+x^2}$$

$$=\frac{1}{2}\frac{x\left(e^{i\theta}+e^{-i\theta}\right)}{1+ix\left(e^{i\theta}-e^{-i\theta}\right)+x^2} \qquad \left[\because \frac{1}{i}=-i\right]$$

$$=\frac{1}{2}\frac{x\left(e^{i\theta}+e^{-i\theta}\right)}{1+ix\left(e^{i\theta}-e^{-i\theta}\right)-i^2x^2e^{i\theta}e^{-i\theta}} \qquad \textbf{(Note)}$$

$$=\frac{1}{2}\frac{x\left(e^{i\theta}+e^{-i\theta}\right)}{1+ix\left(e^{i\theta}-e^{-i\theta}\right)-ixe^{i\theta}.ixe^{-i\theta}}=\frac{1}{2i}\frac{ix\left(e^{i\theta}+e^{-i\theta}\right)}{\left(1+ixe^{i\theta}\right)\left(1-ixe^{-i\theta}\right)}$$

$$=\frac{1}{2i}\left[\frac{1}{1-ixe^{-i\theta}}-\frac{1}{1+ixe^{i\theta}}\right], \qquad \text{(by partial fractions)}$$

$$=\frac{1}{2i}\left[\left(1-ixe^{-i\theta}\right)^{-1}-\left(1+ixe^{i\theta}\right)^{-1}\right].$$

The binomial expansions of

$(1 - ix\ e^{-i\theta})^{-1}$ and $(1 + ix\ e^{i\theta})^{-1}$

are valid since $|ix\ e^{-i\theta}| = |x| < 1$

and similarly $|ix\ e^{i\theta}| = |x| < 1$, x being given to be < 1.

Hence $\dfrac{x\cos\theta}{1-2x\sin\theta+x^2}$

$$=\left(\frac{1}{2i}\right)[(1 + ix\ e^{-i\theta} + i^2x^2\ e^{-2i\theta} + i^3x^3e^{-3i\theta} + ...)$$

$$- (1 - ix\ e^{i\theta} + i^2x^2\ e^{2i\theta} - i^3x^3e^{3i\theta} + ...)]$$

$$=\left(\frac{1}{2i}\right)[(ix\ (e^{i\theta} + e^{-i\theta}) - i^2x^2\ (e^{2i\theta} - e^{-2i\theta})$$

$$+ i^3x^3\ (e^{3i\theta} + e^{-3i\theta}) - i^4x^4\ (e^{4i\theta} - e^{-4i\theta}) + ...]$$

$$=x\left(\frac{e^{i\theta}+e^{-i\theta}}{2}\right)+x^2\left(\frac{e^{2i\theta}-e^{-2i\theta}}{2i}\right)-x^3\left(\frac{e^{3i\theta}+e^{-3i\theta}}{2}\right)-x^4\left(\frac{e^{4i\theta}-e^{-4i\theta}}{2i}\right)+...$$

$$= x\cos\theta + x^2\sin 2\theta - x^3\cos 3\theta - x^4\sin 4\theta + ...$$

6. *Expansion of* $\dfrac{\sin n\theta}{\sin\theta}$ *in a series of descending powers of cos θ.*

in 4 we have proved that

$$\frac{x\sin\theta}{1-2x\cos\theta+x^2} = x\sin\theta + x^2\sin 2\theta + ... + x^n\sin n\theta + ...$$

Dividing both sides by x sin θ, we have

$$\frac{1}{1-2x\cos\theta+x^2}=1+\frac{x\sin 2\theta}{\sin\theta}+\frac{x^2\sin 3\theta}{\sin\theta}+...+\frac{x^{n-1}\sin n\theta}{\sin\theta}+......(1)$$

We observe that $\frac{\sin n\theta}{\sin\theta}$ is the coefficient of x^{n-1} in the R.H.S. of (1).

$\therefore\ \frac{\sin n\theta}{\sin\theta}$ = the coefficient of x^{n-1} in the R.H.S. of (1)

= the coeff. of x^{n-1} in the expansion of $\frac{1}{1-2x\cos\theta+x^2}$

= the coeff. of x^{n-1} in the expansion of $[1 - 2x\cos\theta + x^2]^{-1}$

= the coeff. of x^{n-1} in the expansion of $[1 - x(2\cos\theta - x)]^{-1}$

= the coeff. of x^{n-1} in $[1 + x(2\cos\theta - x) + x^2(2\cos\theta - x)^2$

$+ x^3(2\cos\theta - x)^3 + ... + x^{n-1}(2\cos\theta - x)^{n-1} + ... \infty]$.

Now the terms after $x^{n-1}(2\cos\theta - x)^{n-1}$ contain x^n and higher powers of x which we do not require.

The coeff. of x^{n-1}

in $x^{n-1}(2\cos\theta - x)^{n-1}$ is $(2\cos\theta)^{n-1}$,

in $x^{n-2}(2\cos\theta - x)^{n-2}$ is $-(n-2)(2\cos\theta)^{n-3}$,

in $x^{n-3}(2\cos\theta - x)^{n-3}$ is $\frac{(n-3)(n-4)}{1.2}(2\cos\theta)^{n-5}$,

and so on.

Hence $\frac{\sin n\theta}{\sin\theta} = (2\cos\theta)^{n-1} - (n-2)(2\cos\theta)^{n-3} + \frac{(n-3)(n-4)}{1.2}$ $(2\cos\theta)^{n-5} - ...,$

the last term being $(-1)^{(n-2)/2}(2\cos\theta)$ if n is even

and $(-1)^{(n-1)/2}$ if n is odd.

7. *Expansion of* $\frac{1-x^2}{1-2x\cos\theta+x^2}$ *in a series of cosines of multiples of* θ, $x < 1$.

Or

Expand $\frac{1-x^2}{1-2x\cos\theta+x^2}$ *in a series of ascneding powers of* x.

Here the highest power of x in the numerator and denominator is the same (*i.e.*, 2), therefore we shall divide the Nr. by the Dr. to make the highest power of x in the Nr. les than that in the Dr.

$$\therefore \frac{1-x^2}{1-2x\cos\theta+x^2}=-1+\frac{2-2x\cos\theta}{1-2x\cos\theta+x^2}$$

$$=-1+\frac{2-2x.\frac{1}{2}\left(e^{i\theta}+e^{-i\theta}\right)}{1-2x.\frac{1}{2}\left(e^{i\theta}+e^{-i\theta}\right)+x^2}=-1+\frac{2-x\left(e^{i\theta}+e^{-i\theta}\right)}{\left(1-xe^{i\theta}\right)\left(1-xe^{-i\theta}\right)}$$

$$=-1+\frac{\left(1-xe^{i\theta}\right)+\left(1-xe^{-i\theta}\right)}{\left(1-xe^{i\theta}\right)\left(1-xe^{-i\theta}\right)}=-1+\frac{1}{\left(1-xe^{-i\theta}\right)}+\frac{1}{\left(1-xe^{i\theta}\right)}$$

$= -1 + (1 - xe^{-i\theta})^{-1} + (1 - xe^{-i\theta})^{-1}$

$= -1 + (1 + xe^{-i\theta} + x^2e^{-2i\theta} + ...) + (1 + xe^{i\theta} + x^2e^{2i\theta} + ...)$,

expanding the terms on the R.H.S. by binomial expansion, x being < 1

$= 1 + x\,(e^{i\theta} + e^{-i\theta}) + x^2\,(e^{2i\theta} + e^{-2i\theta}) + ...$

$= 1 + 2x \cos\theta + 2x^2 \cos 2\theta + ... + 2x^n \cos n\theta + ...$ ad.inf.

8. *Expansion of cos nθ in a series of descending powers of cos θ*

In the previous article (*i.e.*, in 7), we have proved that

$$\frac{1-x^2}{1-2x\cos\theta+x^2} = 1 + 2x \cos\theta + 2x^2 \cos 2\theta + ... + 2x^n \cos n\theta + ...$$

$\therefore$ $2 \cos n\theta$ = the coefficient of x^n in the R.H.S.

= the coefficient of x^n in $\dfrac{1-x^2}{1-2x\cos\theta+x^2}$

= the coeffi. of x^n in $(1 - 2x \cos\theta + x^2)^{-1}$

– the coeff. of x^{n-2} in $(1 - 2x \cos\theta + x^2)^{-1}$.

Now $(1 - 2x \cos\theta + x^2)^{-1} = [1 - x\,(2 \cos\theta - x)]^{-1}$

$= 1 + x\,(2 \cos\theta - x) + x^2\,(2 \cos\theta - x)^2 + ...$

$+ x^n\,(2 \cos\theta - x)^n + ...$ ...(1)

From the R.H.S. of (1), the coeff. of x^n

in $x^n\,(2 \cos\theta - x)^n$ is $(2 \cos\theta)^n$,

in $x^{n-1}\,(2 \cos\theta - x)^{n-1}$ is $-(n-1)\,(2 \cos\theta)^{n-2}$,

in $x^{n-2}\,(2 \cos\theta - x)^{n-2}$ is $\dfrac{(n-2)(n-3)}{2!}(2\cos\theta)^{n-4}$, in x^{n-3} $(2 \cos \theta - x)^{n-3}$ is $\dfrac{-(n-3)(n-4)(n-5)}{3!}(3\cos\theta)^{n-6}$,

and so on.

Adding all these we shall get the coefficient of x^n in

$$(1 - 2x \cos \theta + x^2)^{-1}.$$

Again from the R.H.S. of (1), the coeff. of x^{n-2}

in x^{n-2} $(2 \cos \theta - x)^{n-2}$ is $(2 \cos \theta)^{n-2}$,

in x^{n-3} $(2 \cos \theta - x)^{n-3}$ is $-(n-3)(2 \cos \theta)^{n-4}$,

in x^{n-4} $(2 \cos \theta - x)^{n-4}$ is $\dfrac{(n-4)(n-5)}{2!}$ $(2 \cos \theta)^{n-6}$,

and so on.

Adding all these we shall get the coeff. of

$$x^{n-2} \text{ in } (1 - 2x \cos \theta + x^2)^{-1}.$$

Hence $2 \cos n\theta = [(2 \cos \theta)^n - (n-1)(2 \cos \theta)^{n-2}$

$+ \{(n-2)(n-3)/2!\}(2 \cos \theta)^{n-4}$

$- \{(n-3)(n-4)(n-5)/3!\}(2 \cos \theta)^{n-6} + ...]$

$- [(2 \cos \theta)^{n-2} - (n-3)(2 \cos \theta)^{n-4} + \{(n-4)(n-5)/2!\}$

$(2 \cos \theta)^{n-6} - ...]$

$$= (2 \cos \theta)^n - n(2 \cos \theta)^{n-2} + \left[\frac{(n-2)(n-3)}{2!} + (n-3)\right](2 \cos \theta)^{n-4}$$

$$-\left[\frac{(n-3)(n-4)(n-5)}{3!} + \frac{(n-4)(n-5)}{2!}\right](2\cos\theta)^{n-6} + ...$$

$$= (2 \cos \theta)^n - n(2 \cos \theta)^{n-2} + \frac{n(n-3)}{2!}(2\cos\theta)^{n-4}$$

$$-\frac{n(n-4)(n-5)}{3!}(2\cos\theta)^{n-6} + ...$$

The last term being $(-1)^{(n-1)/2}(2 \cos \theta)\, n$, if n is odd and $(-1)^{n/2}.2$, if n is even.

9. *Expansion of log $(1 - 2x \cos \theta + x^2)$ in a series of cosines of multiples of θ.*

We have $\log (1 - 2x \cos \theta + x^2)$

$= \log [1 - x (e^{i\theta} + e^{-i\theta}) + x^2]$ $\left[\because \cos\theta = \frac{1}{2}\left(e^{i\theta} + e^{-i\theta}\right)\right]$

$= \log [(1 - xe^{i\theta})(1 - xe^{-i\theta})] = \log (1 - xe^{i\theta}) + \log (1 - xe^{-i\theta})$

$$= \left(-xe^{i\theta} - \frac{1}{2}x^2e^{2i\theta} - \frac{1}{3}x^3e^{3i\theta} - ...\right)$$

$$+\left(-xe^{-i\theta} - \frac{1}{2}x^2e^{-2i\theta} - \frac{1}{3}x^3e^{-3i\theta} - ...\right),$$

expanding both $\log (1 - xe^{i\theta})$ and $\log (1 - xe^{-i\theta})$ by logarithmic expansion which is justified if $|x| > 1$

$$= - x (e^{i\theta} + e^{-i\theta}) - \frac{1}{2}x^2\left(e^{2i\theta} + e^{-2i\theta}\right) - \frac{1}{3}x^3\left(e^{3i\theta} + e^{-3i\theta}\right) - ...$$

$$= -2\left[x\cos\theta + \frac{1}{2}x^2\cos 2\theta + \frac{1}{3}x^3\cos 3\theta + ...\text{ad.inf.}\right]$$

10. *If sin x = n sin (a + x), expand x in a series of ascending powers of n, where n is less than unity.*

We have $\sin x = n \sin (a + x)$.

$$\therefore \frac{e^{ix} - e^{-ix}}{2i} = n.\frac{e^{i(a+x)} - e^{-i(a+x)}}{2i} \qquad \left(\because \sin\theta = \frac{e^{i\theta} - e^{-i\theta}}{2i}\right)$$

or $(e^{ix} - e^{-ix}) = n [e^{i(a + x)} - e^{-i(a + x)}]$

or $e^{2ix} - 1 = n (e^{ia}.e^{2ix} - e^{-ia})$ [Multiplying both sides by e^{ix}]

or $e^{2ix} (1 - ne^{ia}) = (1 - ne^{-ia})$

or $e^{2ix} = \{(1 - ne^{-na})/(1 - ne^{ia})\}$.

Taking logarithm of both sides, we have

$2ix = \log (1 - ne^{-ia}) - \log (1 - ne^{ia})$

$$= \left(-ne^{-ia} - \frac{1}{2}n^2e^{-2ia} - \frac{1}{3}n^3e^{-3ia} - ...\right)$$

$$-\left(-ne^{ia} - \frac{1}{2}n^2e^{2ia} - \frac{1}{3}n^3e^{3ia} - ...\right),$$

the logarithmic expansions are valid because $|ne^{-ia}| = |n| < 1$ and also $|ne^{ia}| = |n| < 1$.

$$= n\left(e^{ia} - e^{-ia}\right) + \frac{1}{2}n^2\left(e^{2ia} - e^{-2ia}\right) + \frac{1}{3}n^3\left(e^{3ia} - e^{-3ia}\right) + ...$$

$$\therefore \quad x = n\left(\frac{e^{ia} - e^{-ia}}{2i}\right) + \frac{1}{2}n^2\left(\frac{e^{2ia} - e^{-2ia}}{2i}\right) + \frac{1}{3}n^3\left(\frac{e^{3ia} - e^{-3ia}}{2i}\right) + \dots$$

$$= n \sin a + \frac{1}{2}n^2 \sin 2a + \frac{1}{3}n^3 \sin 3a + \dots$$

SUMMATION OF TRIGONOMETRICAL SERIES

In this method we shall deal with the summation of some trigonometrical series with a finite and infinite number of terms. It may be noted that all the series with infinite number of terms are not summable and hence due care should be taken regarding the convergence of such series.

GENERAL C + IS METHOD

A general method for the Summation of Trigonometrical Series of the type:

$$C_1 \cos \theta + C_2 \cos (\theta + \alpha) + C_3 \cos (\theta + 2\alpha) +$$

Or $C_1 \sin \theta + C_2 \sin (\theta + \alpha) + C_3 \sin (\theta + 2\alpha) +$

is to denote the serie (1) by C, since the coefficients of C's in this series are cosines of the angles increasing in arithmetic progression and the series (2) by S being a sine series. Multiplying (2) by i and adding

$$C + is = C_1 (\cos \theta + i \sin \theta) + C_2 \{\cos (\theta + \alpha) + i \sin (\theta + \alpha)\} + C_3 \cos (\theta + 2\alpha) + i \sin (\theta + 2\alpha) +$$

$$= C_1 e^i \theta + C_2 e^i (\theta + \alpha) C_3 e^i (\theta + 2\alpha) +$$

$$= e^i \theta \{C_1 + C_1 e^{i\alpha} + C_3 e^{2i\alpha} + \dots\}$$

Now the series within the bracket of (3) may be either:

(i) Series in Geometrical progression.

(ii) Binomial series or one which can be reduced to it.

(iii) Exponential series or the allied series *e.g.*, sine or cosine series, and

(iv) Logarithmic series or the allied series *e.g.*, Gregory's series.

The real part of (3) gives C and the imaginary part gives S. If either of the C or S is given, the other series known as auxilliary series can be formed and the sum of C+ is found the real part of the sum. So found is equal to C and the imaginary part equal to so.

USE OF GEOMATRIC SERIES

The geomatric series is given by

$$a + az^2 + az^2 + az^3 + \dots a_n \text{ inf}$$

sums of n terms and infinity of the above series are given by

$$s_n = \frac{a(z^n - 1)}{z-1} \text{ if } |z| > 1$$

$$= \frac{a(1-z^n)}{1-z} \text{ if } |z| < 1$$

$$s_\infty = \frac{a}{1-z} \text{ provided } |z| < 1.$$

SUMMATION OF SERIES

The following formulae are very useful in solving problems of Summation of trigonometrical series:

1. $e^{ix} = \cos x + i \sin x.$ **[Euler's formula]**
2. $e^{-ix} = \cos x - i \sin x.$
3. $e^{ix} + e^{-ix} = 2 \cos x$, $e^{ix} - e^{-ix} = 2i \sin x.$
4. $e^{inx} = \cos nx + i \sin nx$, n is an integer.
5. $(1 + a)^n = 1 + na + \frac{n(n-1)}{2!} a^2 + \frac{n(n-1)(n-2)}{3!} a^3 + ... + a^n$
6. $e^x = 1 + x + \frac{x^2}{2!} + \frac{x^3}{3!} + ...$
7. $1 + x + x^2 + ... + x^{n-1} = \frac{1-x^n}{1-x}, x \neq 1.$

The following rule is used in the Summation of Trigonometrical Series:

Rule: *Write the cosines series as C and the sine series as S. Simplify C + iS use the above expansions.*

Example 1:

Find the sum of series:

$$\cos\theta - \frac{1}{2}\cos 2\theta + \frac{1}{3}\cos 3\theta - ... \text{ ad inf.}$$

$$\sin\theta - \frac{1}{2}\sin 2\theta + \frac{1}{3}\sin 3\theta - ... \text{ ad inf.}$$

Solution:

Let $C = \cos\theta - \frac{1}{2}\cos 2\theta + \frac{1}{3}\cos 3\theta - .....$

and $S = \sin\theta - \frac{1}{2}\sin 2\theta + \frac{1}{3}\sin 3\theta - ....$

$\therefore$ $C + iS = (\cos\theta + i\sin\theta) - \frac{1}{2}(\cos 2\theta + i\sin 2\theta) + ...$

$= (\cos\theta + i\sin\theta) - \frac{1}{2}(\cos\theta + i\sin\theta)^2 + \frac{1}{3}(\cos\theta + i\sin\theta)^3 - .....$

$= z - \frac{1}{2}z^2 + \frac{1}{3}z^3 - ....,\ z = \cos\theta + i\sin\theta$

$= \log(1 + z)$

$= \log(1 + \cos\theta + i\sin\theta)$

$= \log(2\cos^2\theta/2 + 2i\sin\theta/2\cos\theta/2)$

$= \log[2\cos\theta/2(\cos\theta/2 + i\sin\theta/2)]$

$= \log[2\cos\theta/2.e^{i\theta/2}]$

$= \log(2\cos\theta/2) + \log(e^{i\theta/2})$

$= \log(2\cos\theta/2) + i\theta/2.$

Comparing real and imaginary parts on both sides, we get

$C = \log(2\cos\theta/2),$

$S = \frac{1}{2}\theta.$

Example 2:

Sum the series:

$\sin\alpha + c\sin(\alpha + \beta) + \frac{c^2}{2!}\sin(\alpha + 2\beta) + ... + ad.\ inf.$

Solution:

Let $S = \sin\alpha + c\sin(\alpha + \beta) + \frac{c^2}{2!}\sin(\alpha + 2\beta) + ...\infty$

and $C = \cos\alpha + c\cos(\alpha + \beta) + \frac{c^2}{2!}\cos(\alpha + 2\beta) + ...\infty$

$\therefore\ C + iS = (\cos\alpha + i\sin\alpha) + c\{\cos(\alpha + \beta) + i\sin(\alpha + \beta)\}$

$+ \frac{c^2}{2!}\{\cos(\alpha + 2\beta) + i\sin(\alpha + 2\beta)\} + ...\infty$

$= e^{i\alpha} + c\,e^{i(\alpha+\beta)} + \frac{c^2}{2!}e^{i(\alpha+2\beta)} + ...\infty$

$= e^{i\alpha}\left[i + c\,e^{i\beta} + \frac{c^2}{2!}e^{2i\beta} + ...\infty\right]$

$$= e^{i\alpha}\left[i + cz + \frac{1}{2\,i}c^2z^2 + \frac{1}{3\,i}c^3z^3 + \ldots\infty\right]_{,\,z\,=\,e^{i\beta}}$$

$$= e^{i\alpha}e^{cz} = (\cos\alpha + i\sin\alpha)\,e^{c(\cos\beta + i\sin\beta)}$$

$$= e^{c\cos\beta}(\cos\alpha + i\sin\alpha)\,e^{ic\,\sin\beta}$$

$$C + iS = e^{c\cos\beta}(\cos\alpha + i\sin\alpha)\{\cos(c\sin\beta) + i\sin(c\sin\beta)\}$$

Equating imaginary parts on both sides, we obtain

$$S = e^{c\cos\beta}\{\sin\alpha\cos(c\sin\beta) + \cos\alpha\sin(c\sin\beta)\}$$

Hence $S = e^{c\cos\beta}\sin(\alpha + c\sin\beta)$.

Example 3:

Prove that if n is a positive integer, then

(i) $\cos\theta + {}^nc_1\cos(\theta + \phi) + {}^nc_1\cos(\theta + 2\phi) + \ldots + {}^nc_n\cos(\theta + n\phi)$

$$= (2\cos\phi/2)^n\cos\left(\theta + \frac{1}{2}n\phi\right).$$

(ii) $\sin\theta + {}^nc_1\sin(\theta + \phi) + {}^nc_2\sin(\theta + 2\phi) + \ldots + {}^nc_n\sin(\theta + n\phi)$

$$= (2\cos\phi/2)^n\sin\left(\theta + \frac{1}{2}n\phi\right).$$

Solution:

Let $C = \cos\theta + {}^nc_1\cos(\theta + \phi) + {}^nc_2\cos(\theta + 2\phi)$

$+ \ldots + {}^nc_n\sin(\theta + n\phi)$.

$$\therefore\quad C + iS = e^{i\theta} + {}^nc_1\,e^{i(\theta+\phi)} + {}^nc_2\,e^{i(\theta - 2\phi)} + \ldots + {}^nc_n\,e^{i(\theta+n\phi)}.$$

$$= e^{i\theta}[1 + {}^nc_1\,e^{i\phi} + {}^nc_2\,e^{i2\phi} + \ldots + {}^nc_n\,e^{in\phi}]$$

$$= e^{i\theta}(1 + e^{i\phi})^n$$

$$= e^{i\theta}[1 + \cos\phi + i\sin\phi]^n$$

$$= e^{i\theta}[2\cos^2\phi/2 + 2i\sin\phi/2\cos\phi/2]^n$$

$$e^{i\theta}(2\cos\phi/2)^n[\cos\phi/2 + i\sin\phi/2]^n$$

$$\therefore\quad C + iS = (2\cos\phi/2)^n(\cos\theta + i\sin\theta)(\cos n\phi/2 + i\sin n\phi/2) \quad \ldots(1)$$

Comparing real parts on both sides,we get

$$C = (2\cos\phi/2)^n[\cos\theta\cos n\phi/2 - \sin\theta\sin n\phi/2]$$

Hence $$C = (2\cos\phi/2)^n\cos\left(\theta + \frac{1}{2}n\phi\right).$$

Similarly on comparing imaginary parts on both sies of (1),

$$S = (2 \cos \phi/2)^n \sin \left(\theta + \frac{1}{2} n\phi\right).$$

Example 4:

Sum to n terms:

$$n \cos \theta + \frac{n(n-1)}{1.2} \cos 2\theta + \frac{n(n-1)(n-2)}{1.2.3} \cos 3\theta + ...$$

Solution:

Let

$$C = n \cos \theta + \frac{n(n-1)}{1.2} \cos 2\theta + \frac{n(n-1)(n-2)}{1.2.3} \cos 3\theta + ... + \cos n\theta,$$

$$S = n \sin \theta + \frac{n(n-1)}{1.2} \cos 2\theta + \frac{n(n-1)(n-2)}{1.2.3} \sin 3\theta + ... + \sin n\theta.$$

$$\therefore \quad C + iS = \left[1 + ne^{i\theta} + \frac{n(n-1)}{1.2} e^{12\theta} + ... + e^{in\theta}\right] - 1$$

$$= (1 + e^{i\theta})^n - 1$$

$$= (1 + \cos^2 \theta + i \sin \theta)^n - 1$$

$$= (2 \cos^2 \theta/2 + 2i \sin \theta/2 \cos \theta/2)^n - 1$$

$$(2 \cos \theta/2)^n [\cos \theta/2 + i \sin \theta/2]^n - 1$$

$$= (2 \cos \theta/2)^n [\cos (n\theta/2) + i \sin (n\theta/2)] - 1.$$

Comparing real parts on both sides, we get

$$C = (2 \cos \theta/2)^n \cos (n\theta/2) - 1.$$

Example 5:

Sum to n terms:

(1) $\cos \theta + \cos 2\theta + \cos 3\theta + ...$

(2) $\sin \theta + \sin 2\theta + \sin 3\theta + ...$

Solution:

Let $C = \cos \theta + \cos 2\theta + ... + \cos n\theta$,

and $S = \sin \theta + \sin 2\theta + ... + \sin n\theta$.

$\therefore C + iS = (\cos \theta + i \sin \theta) + (\cos 2\theta + i \sin 2\theta) + ...$

$$+ (\cos n\theta + i \sin n\theta)$$

$$= (\cos\theta + i\sin\theta) + (\cos\theta + i\sin\theta)^2 + \ldots$$

$$+ (\cos\theta + i\sin\theta)^n \quad \text{[by DeMoivre's Theorem]}$$

$$= z + z^2 + \ldots + z^n, \text{ where } z = \cos\theta + i\sin\theta.$$

$$= z(1 + z + z^2 + \ldots + z^{n-1}) = \frac{z(1-z^n)}{1-z}$$

$$= \frac{(\cos\theta + i\sin\theta)\left\{1 - (\cos\theta + i\sin\theta)^n\right\}}{1 - \cos\theta - i\sin\theta}$$

$$= \frac{(\cos\theta + i\sin\theta) - (\cos\theta + i\sin\theta)^{n+1}}{(1 - \cos\theta) - i\sin\theta} \times \frac{(1-\cos\theta) + i\sin\theta}{(1-\cos\theta) + i\sin\theta}$$

$$= (\cos\theta + i\sin\theta) - \{\cos(n+1)\theta + i\sin(n+1)\theta\}$$

$$\times \frac{(1-\cos\theta) + i\sin\theta}{(1-\cos\theta)^2 + i\sin^2\theta}$$

$$= \frac{\left[\{\cos\theta - \cos(n+1)\theta\} + i\{\sin\theta - \sin(n+1)\theta\}\right]\{(1-\cos\theta) + i\sin\theta\}}{2 - 2\cos\theta}$$

$$= \frac{\begin{array}{l}\left[\{\cos\theta - \cos(n+1)\theta\}(1-\cos\theta) - \{\sin\theta - \sin(n+1)\theta\}\sin\theta\right] \\ +i\left[\sin\theta\{\cos\theta - \cos(n+1)\theta\} + \{\sin\theta - \sin(n+1)\theta\}(1-\cos\theta)\right]\end{array}}{2(1-\cos\theta)}$$

$$= \frac{\begin{array}{l}\left[\cos\theta\{\cos(n+1)\theta\cos\theta + \sin(n+1)\theta\sin\theta\} - \cos(n+1)\theta - 1\right] \\ +i\left[\sin\theta + \{\sin(n+1)\theta\cos\theta - \cos(n+1)\theta\sin\theta\} - \sin(n+1)\theta\right]\end{array}}{2(1-\cos\theta)}$$

Equating real and imaginary parts on both sides, we get

$$C = \frac{\cos\theta + \cos n\theta - \cos(n+1)\theta - 1}{2(1-\cos\theta)}$$

$$S = \frac{\sin\theta + \sin n\theta - \sin(n+1)\theta}{2(1-\cos\theta)}.$$

Example 6:

Sum to n terms

(i) $\cos\alpha + \cos(\alpha+\beta) + \cos(\alpha+2\beta) + \ldots$

(ii) $\sin\alpha + \sin(\alpha+\beta) + \sin(\alpha+2\beta) + \ldots, \beta \neq 2k\pi.$

Solution:

Let $C = \cos\alpha + \cos(\alpha+\beta) + \ldots + \cos\{\alpha + (n-1)\beta\}$,

and $S = \sin\alpha + \sin(\alpha+\beta) + \ldots + \sin\{\alpha + (n-1)\beta\}$.

$$\therefore C + iS = (\cos\alpha + i\sin\alpha) + \{\cos(\alpha+\beta) + i\sin(\alpha+\beta)\} + \ldots + [\cos\{\alpha+(n-1)\beta\} + i\sin\{\alpha+(n-1)\beta\}]$$

$$= e^{i\alpha} + e^{i(\alpha+\beta)} + e^{i(\alpha+2\beta)} + \ldots + e^{i[\alpha+(n-1)\beta]}$$

$$= e^{i\alpha}\{1 + e^{i\beta} + e^{2i\beta} + \ldots + e^{(n-1)i\beta}\}$$

$$= e^{i\alpha}\{1 + z + z^2 + \ldots + z^{n-1}\},\ z = e^{i\beta} = \cos\beta + i\sin\beta$$

$$= e^{ia}\left(\frac{1-z^n}{1-z}\right)$$

$$= \frac{e^{i\alpha}(1-\cos n\beta - i\sin n\beta)}{1-\cos\beta - i\sin\beta}$$

$$= \frac{e^{i\alpha}\{2\sin^2(n\beta/2) - 2i\sin(n\beta/2)\cos(n\beta/2)\}}{2\sin^2(\beta/2) - 2i\sin(\beta/2)\cos(\beta/2)}$$

$$= \frac{-2i\,e^{i\alpha}\sin(n\beta/2)\{\cos(n\beta/2) + i\sin(n\beta/2)\}}{-2i\sin(\beta/2)\{\cos(\beta/2) + i\sin(\beta/2)\}}$$

$$= \frac{\sin(n\beta/2)}{\sin(\beta/2)}(\cos\alpha + i\sin\alpha)\{\cos(n-1)(\beta/2) + i\sin(n-1)(\beta/2)\}$$

$$= \frac{\sin(n\beta/2)}{\sin(\beta/2)}[\cos\{\alpha + (n-1)(\beta/2)\} + i\sin\{\alpha + (n-1)(\beta/2)\}]$$

Equating real and imaginary parts on both sides,

$$C = \frac{\sin(n\beta/2)}{\sin(\beta/2)}\cos\{\alpha + (n-1)(\beta/2)\}$$

$$S = \frac{\sin(n\beta/2)}{\sin(\beta/2)}\sin\{\alpha + (n-1)(\beta/2)\}.$$

Example 7:

Sum to n terms:

$$\cos\theta + \cos(\theta + \pi/n) + \cos(\theta + 2\pi/n) + \ldots$$

Solution:

Let

$C = \cos\theta + \cos(\theta + \pi/n) + \cos(\theta + 2\pi/n) + ... + \cos\{\theta + (n-1)\pi/n\}$,

$S = \sin\theta + \sin(\theta + \pi/n) + \sin(\theta + 2\pi/n) + ... + \sin\{\theta + (n-1)\pi/n\}$.

$\therefore\ C + iS = e^{i\theta} + e^{i(\theta + \pi/n)} + e^{i(\theta + 2\pi)} + ...$

$= e^{i\theta}[1 + e^{i\pi/n} + e^{i2\pi/n} + ... + e^{i(n-1)\pi/n}]$

$$= \frac{e^{i\theta}(1 - z^n)}{1 - z} = \frac{e^{i\theta}(1 - e^{i\pi})}{1 - e^{i\pi/n}}$$

$$= \frac{e^{i\theta}[1 - (\cos\pi + i\sin\pi)]}{1 - (\cos\pi/n + i\sin\pi/n)}$$

$$= \frac{2e^{i\theta}}{1 - \{1 - 2\sin^2(\pi/2n) + 2i\sin(\pi/2n)\cos(\pi/2n)\}}$$

$$= \frac{e^{i\theta}}{-i\sin(\pi/2n)[\cos(\pi/2n) + i\sin(\pi/2n)]}$$

$$= \frac{ie^{i\theta}(\pi/2n) - i\sin(\pi/2n)}{-i^2\sin(\pi/2n)}$$

$= i(\cos\theta + i\sin\theta)[\cos(\pi/2n) - i\sin(\pi/2n)]\operatorname{cosec}(\pi/2n)$.

Equating the real parts on both the sides, we get

$C = [\cos\theta\sin(\pi/2n) - \sin\theta\cos(\pi/2n)]\operatorname{cosec}(\pi/2n)$.

Hence $C = \sin\left(\frac{\pi}{2n} - \theta\right)\operatorname{cosec}\left(\frac{\pi}{2n}\right)$.

Example 8:

Sum the series

(i) $\cos\theta + x\cos 2\theta + x^2\cos 3\theta + ...$ *n terms.*

(ii) $\sin\theta + x\sin 2\theta + x^2\sin 3\theta + ...$ *n terms.*

Solution:

Let $C = \cos\theta + x\cos 2\theta + x^2\cos 3\theta + ... + x^{n-1}\cos n\theta$.

and $S = \sin\theta + x\sin 2\theta + x^2\sin 3\theta + ... + x^{n-1}\sin n\theta$.

$\therefore\ C + iS = e^{i\theta} + xe^{2i\theta} + x^2e^{3i\theta} + ... + x^{n-1}e^{ni\theta}$

$$= e^{i\theta}[1 + xe^{i\theta} + x^2e^{3i\theta} + ... + x^{n-1}e^{(n-1)\theta}]$$

$$= e^{i\theta}(1 + z + z^2 + ... + z^{n-1}), \text{ where } z = xe^{i\theta}$$

$$= e^{i\theta}\left\{\frac{1-z^n}{1-z}\right\} = e^{i\theta}\left\{\frac{1-x^n e^{in\theta}}{1-xe^{i\theta}}\right\}.$$

$$= e^{i\theta}\left\{\frac{1-x^n e^{in\theta}}{1-xe^{i\theta}} \times \frac{1-xe^{-i\theta}}{1-xe^{-i\theta}}\right\}$$

$$= e^{i\theta}\left[\frac{1-xe^{-i\theta}-x^n e^{in\theta}+x^{n+1}e^{i(n-1)\theta}}{1-x\left(e^{i\theta}+e^{-i\theta}\right)+x^2}\right]$$

$$= \frac{e^{i\theta}-x-x^n e^{i(n+1)\theta}+x^{n+1}e^{in\theta}}{1-2x\cos\theta+x^2}$$

$$= \frac{(\cos\theta+i\sin\theta)-x-x^n\{\cos(n+1)\theta+i\sin(n+1)\theta\} + x^{n+1}\{\cos n\theta+i\sin n\theta\}}{1-2x\cos\theta+x^2}$$

Comparing the real and imaginary parts, we get

$$C = \frac{\cos\theta-x-x^n\cos(n+1)\theta+x^{n+1}\cos n\theta}{1-2x\cos\theta+x^2},$$

$$S = \frac{\sin\theta-x^n\sin(n+1)\theta+x^{n+1}\sin n\theta}{1-2x\cos\theta+x^2}.$$

Example 9:

Sum the n terms

(i) $\cos^2\theta + \cos^2 2\theta + \cos^2 3\theta + ...$

(ii) $\sin^2\theta + \sin^2 2\theta + \sin^2 3\theta + ...$

Solution:

(i) Let $C = \cos^2\theta + \cos^2 2\theta + ... + \cos^2 n\theta$

$$= \frac{1}{2}(1 + \cos 2\theta) + \frac{1}{2}(1 + \cos 4\theta) + ... + \frac{1}{2}(1 + \cos 2n\theta)$$

$$= \frac{n}{2} + \frac{1}{2}(\cos 2\theta + \cos 4\theta + ... + \cos 2n\theta)$$

$$= \frac{n}{2} + \frac{1}{2}(\cos\phi + \cos 2\phi + ... + \cos n\phi), \phi = 2\theta$$

$$= \frac{n}{2} + \frac{1}{2}\left[\frac{\cos\phi - 1 - \cos(n+1)\phi + \cos n\phi}{2(1-\cos\phi)}\right]$$

Hence $C = \dfrac{n}{2} + \dfrac{\cos 2\theta - 1 - \cos 2(n+1)\theta + \cos 2n\theta}{4(1-\cos 2\theta)}$, if $\theta \neq 0$

$= n$ if $\theta = 0$.

(ii) Let $S = \sin^2\theta + \sin^2 2\theta + \sin^2 3\theta + \ldots + \sin^2 n\theta$.

$$= \frac{1}{2}(1 - \cos 2\theta) + \frac{1}{2}(1 - \cos 4\theta) + \ldots + \frac{1}{2}(1 - \cos 2n\theta)$$

$$= \frac{n}{2} - \frac{1}{2}(\cos 2\theta + \cos 4\theta + \ldots + \cos 2n\theta)$$

Hence $S = \dfrac{n}{2} - \dfrac{\cos 2\theta - 1 - \cos 2(n+1)\theta + \cos 2n\theta}{4(1-\cos 2\theta)}$, if $\theta \neq 0$

$= 0$, if $\theta = 0$.

Example 10:

Sum the series

(i) $\cos\theta\cos\theta + \cos^2\theta\cos 2\theta + \cos^3\theta\cos 3\theta + \ldots$ *to n terms.*

(ii) $\cos\theta\sin\theta + \cos^2\theta\sin 2\theta + \cos^3\theta\sin 3\theta + \ldots$ *to n terms.*

Solution:

Let

$C = \cos\theta\cos\theta + \cos^2\theta\cos 2\theta + \ldots + \cos^n\theta\cos n\theta,$

$S = \cos\theta\sin\theta + \cos^2\theta\sin 2\theta + \ldots + \cos^n\theta\sin n\theta.$

$\therefore C + iS = \cos\theta(\cos\theta + i\sin\theta) + \cos^2\theta(\cos 2\theta + i\sin 2\theta) +$

$+ \ldots + \cos^n\theta(\cos n\theta + i\sin n\theta)$

$= z\cos\theta + z^2\cos^2\theta + \ldots + z^n\cos^n\theta,$

where $z = \cos\theta + i\sin\theta$

$= z\cos\theta(1 + z\cos\theta + \ldots + z^{n-1}\cos^{n-1}\theta)$

$$= \frac{z\cos\theta[1 - z^n\cos^n\theta]}{1 - z\cos\theta}$$

$$= \frac{\cos\theta(\cos\theta + i\sin\theta)][1 - \cos^n\theta\{\cos n\theta + i\sin n\theta\}]}{1 - \cos\theta\{\cos\theta + i\sin\theta\}}$$

$$= \frac{\cos\theta(\cos\theta + i\sin\theta)\left[1-\cos^n\theta\cos n\theta - i\cos^n\theta\sin n\theta\right]}{(1-\cos^2\theta) - i\sin\theta\cos\theta}$$

$$= \frac{\cos\theta(\cos\theta + i\sin\theta)\left[1-\cos^n\theta\cos n\theta - i\cos^n\theta\sin n\theta\right]}{-i\sin\theta\,(\cos\theta + i\sin\theta)}$$

$$= i\cot\theta\,[1 - \cos^n\theta\cos n\theta - i\cos^n\theta\sin n\theta]$$

$$= i\cot\theta\,[1 - \cos^n\theta\cos n\theta] + \cos^{n+1}\theta\,\mathrm{cosec}\,\theta\sin n\theta.$$

Equating the real and imaginary parts on both the sides, we get

$$C = \cos^{n+1}\theta\,\mathrm{cosec}\,\theta\sin n\theta,\ \text{if } \theta \neq 0$$

and $C = n$, if $\theta = 0$;

$$S = \cot\theta\,(1 - \cos^n\theta\cos n\theta),\ \text{if } \theta \neq 0$$

and $S = 0$, if $\theta = 0$.

Example 11:

Sum to n terms:

(i) $\cos\theta\cos 2\theta + \cos 2\theta\cos 3\theta + \cos 3\theta\cos 4\theta + \ldots$

(ii) $\sin\theta\sin 2\theta + \sin 2\theta\sin 3\theta + \sin 3\theta\sin 4\theta + \ldots$

Solution:

$$C = \cos\theta\cos 2\theta + \cos 2\theta\cos 3\theta + \ldots + \cos n\theta\cos(n+1)\theta$$

$$= \frac{1}{2}(\cos 3\theta + \cos\theta) + \frac{1}{2}(\cos 5\theta + \cos\theta) + \ldots + \frac{1}{2}[\cos(2n+1)\theta + \cos\theta]$$

$$= \frac{1}{2}[\cos 3\theta + \cos 5\theta + \ldots + \cos(2n+1)] + \frac{n}{2}\cos\theta$$

$$= \frac{1}{2}C_1 + \frac{n}{2}\cos\theta, \qquad \ldots(1)$$

where $C_1 \cos = 3\theta + \cos 5\theta + \ldots + \cos(2n+1)\theta$.

Multiplying both sides by $2\sin\theta$, we get

$$2\sin\theta\,C_1 = 2\sin\theta\cos 3\theta + 2\sin\theta\cos 5\theta + \ldots + 2\sin\theta\cos(2n+1)\theta$$

$$= (\sin 4\theta - \sin 2\theta) + (\sin 6\theta - \sin 4\theta) + \ldots + [\sin(2n+2)\theta - \sin 2n\theta]$$

$= \sin(2n+2)\theta - \sin 2\theta$

$= 2\cos(n+2)\theta \sin n\theta$

$$\therefore \quad C_1 = \frac{\cos(n+2)\theta \sin n\theta}{\sin\theta} \qquad ...(2)$$

Substituting the value of C_1 from (2) in (1), we obtain

$$C = \frac{\cos(n+2)\theta \sin n\theta}{2\sin\theta} + \frac{n}{2}\cos\theta$$

$$= \frac{1}{2}[n\cos\theta + \cos(n+2)\theta \sin n\theta \operatorname{cosec}\theta],\ \theta \neq k\pi.$$

Clearly $C = n$ if θ is an even multiple of π,

$= -n$ if θ is an odd multiple of π.

(ii) Let $S = \sin\theta \sin 2\theta \sin 3\theta + ... + \sin n\theta \sin(n+1)\theta$

$$= \frac{1}{2}(\cos\theta - \cos 3\theta) + \frac{1}{2}(\cos\theta - \cos 5\theta) + ... + \frac{1}{2}[\cos\theta - \cos(2n+1)\theta]$$

$$= \frac{n}{2}\cos\theta - \frac{1}{2}[\cos 3\theta + \cos 5\theta + ... + \cos(2n+1)\theta]$$

$$= \frac{n}{2}\cos\theta - \frac{1}{2}C_1.$$

Hence, on using (2),

$$S = \frac{1}{2}[n\cos\theta - \cos(n+2)\theta \sin n\theta \operatorname{cosec}\theta]\ \theta \neq k\pi$$

$= 0$, if $\theta = k\pi$.

Example 12:

Sum to n terms:

$\sin\theta \cos 2\theta + \sin 2\theta \cos 3\theta + \sin 3\theta \cos 4\theta + ...$

Solution:

Let $A = \sin\theta\cos 2\theta + \sin 2\theta \cos 3\theta + ... + \sin n\theta \cos(n+1)\theta$

$$= \frac{1}{2}(\sin 3\theta - \sin\theta) + \frac{1}{2}(\sin 5\theta - \sin\theta) + ... + \frac{1}{2}[\sin(2n+1)\theta - \sin\theta]$$

$$= \frac{1}{2}[\sin 3\theta + \sin 5\theta + ... + \sin(2n+1)\theta] - \frac{n}{2}\sin\theta$$

$$= \frac{1}{2} S - \frac{n}{2} \sin\theta. \qquad ...(3)$$

where $S = \sin 3\theta + \sin 5\theta + ... + \sin (2n + 1) \theta$

Multiplying both sides by $2 \sin \theta$, we get

$$2 \sin\theta\, S = 2 \sin\theta \sin 3\theta + 2 \sin\theta \sin 5\theta + ... + 2 \sin\theta \sin (2n + 1) \theta$$

$$= (\cos 2\theta - \cos 4\theta) + (\cos 4\theta - \cos 6\theta) + ... + [\cos 2n\theta - \cos (2n + 2) \theta]$$

$$= \cos 2\theta - \cos (2n + 2) \theta$$

$$= 2 \sin (n + 2) \theta \sin n\theta$$

$\therefore \quad S = \sin (n + 2) \theta \sin n\theta \operatorname{cosec} \theta.$

Substituting for S in (3), we get

$$A = \frac{1}{2} [\sin (n + 2) \theta \sin n\theta \operatorname{cosec} \theta - n \sin \theta], \text{ if } \theta \neq k\pi$$

$$= 0, \text{ if } \theta = k\pi.$$

Example 13:

Sum the series

(i) $\cos\theta + \cos 3\theta + \cos 5\theta + ...$ *to n terms.*

(ii) $\sin\theta + \sin 3\theta + \sin 5\theta + ...$ *to n terms.*

Solution:

(i) Let $C = \cos\theta + \cos 3\theta + \cos 5\theta + ... + \cos (2n - 1) \theta.$

$$\therefore 2 \sin\theta\, C = 2 \sin\theta \cos\theta + 2 \sin\theta \cos 3\theta + 2 \sin\theta \cos 5\theta + ... + 2 \sin\theta \cos (2n - 1) \theta$$

$$= \sin 2\theta + (\sin 4\theta - \sin 2\theta) + (\sin 6\theta - \sin 4\theta) + ... + [\sin 2n\theta - \sin (2n ...2) \theta]$$

$$= \sin 2n\theta.$$

Hence $C = \dfrac{\sin 2n\theta}{2 \sin\theta}.$

(ii) Let $S = \sin\theta + \sin 3\theta + \sin 5\theta + ... + \sin (2n - 1) \theta$

$$\therefore \quad 2 \sin\theta\, S = 2 \sin^2\theta + 2 \sin 3\theta \sin\theta + 2 \sin 5\theta \sin\theta + ... + 2 \sin (2n - 1) \theta \sin\theta$$

$$= 2\sin^2\theta + (\cos 2\theta - \cos 4\theta) + (\cos 4\theta - \cos 6\theta) + \ldots$$
$$+ [\cos (2n-2)\theta - \cos 2n\theta\}$$
$$= 2\sin^2\theta + \cos 2\theta - \cos 2n\theta$$
$$= 2\sin^2\theta + 2\cos^2\theta - 1 - \{1 - 2\sin^2 n\theta\}$$
$$= 2\sin^2 n\theta.$$

Hence $$S = \frac{\sin^2 n\theta}{\sin\theta}.$$

Example 14:

Sum the series

$$\cos\alpha + n\cos 2\alpha + \frac{n(n-1)}{2!}\cos 3\alpha + \ldots \text{ upto } (n+1) \text{ terms.}$$

Solution:

Let $$C = \cos\alpha + n\cos 2\alpha + \frac{n(n-1)}{2!}\cos 3\alpha + \ldots.$$

and $$S = \sin\alpha + n\sin 2\alpha + \frac{n(n-1)}{2!}\sin 3\alpha + \ldots.$$

$$\therefore \quad C + iS = (\cos\alpha + i\sin\alpha) + n(\cos 2\alpha + i\sin 2\alpha) + \ldots$$
$$= e^{i\alpha} + ne^{i2\alpha} + \frac{n(n-1)}{2!}e^{i2\alpha} + \ldots \text{ to } (n+1) \text{ terms}$$
$$\Rightarrow \quad C + iS = e^{i\alpha}\left[1 + ne^{i\alpha} + \frac{n(n-1)}{2!}e^{i2\alpha} + \ldots e^{in\alpha}\right]$$
$$= e^{i\alpha}(1 + e^{i\alpha})^n$$
$$= (\cos\alpha + i\sin\alpha)(1 + \cos\alpha + i\sin\alpha)^n$$
$$= (\cos\alpha + i\sin\alpha)\left(2\cos^2\frac{\alpha}{2} + 2i\sin\frac{\alpha}{2}\cos\frac{\alpha}{2}\right)^n$$
$$= 2^n\cos^n\frac{\alpha}{2}(\cos\alpha + i\sin\alpha)\left(\cos\frac{\alpha}{2} + i\sin\frac{\alpha}{2}\right)^n$$
$$= 2^n\cos^n\frac{\alpha}{2}(\cos\alpha + i\sin\alpha)\left(\cos\frac{n\alpha}{2} + i\sin\frac{n\alpha}{2}\right)$$

Comparing the real parts on both the sides, we get

$$C = 2^n \cos^n \frac{\alpha}{2}\left[\cos\alpha\cos\frac{n\alpha}{2} - \sin\alpha\sin\frac{n\alpha}{2}\right]$$

$$= 2^n \cos^n \frac{\alpha}{2}\cos\left(\alpha + \frac{n\alpha}{2}\right)$$

Hence $C = 2^n \cos^n \dfrac{\alpha}{2}\cos\dfrac{(n+2)\,\alpha}{2}$.

Example 15:

Apply Summation of Series to prove that the equation whose roots are nth powers of the roots of the the equation $x^2 - 2\cos\theta\, x + 1 = 0$ is $x^2 - 2\cos n\theta\, x + 1 = 0$.

Solution:

Solving the equation $x^2 - 2\cos\theta\, x + 1 = 0$, we get

$$x = \frac{2\cos\theta \pm \sqrt{4\cos^2\theta - 4}}{2} = \cos\theta \pm i\sin\theta.$$

The nth powers of the above two roots are

$\alpha = (\cos\theta + i\sin\theta)^n,\ \beta = (\cos\theta - i\sin\theta)^n$

$\Rightarrow$ $\alpha = \cos n\theta + i\sin n\theta,\ \beta = \cos n\theta - i\sin n\theta$, by DeMoivre's theorem

$\therefore$ $\alpha + \beta = 2\cos n\theta,\ \alpha\beta = \cos^2 n\theta + \sin^2 n\theta = 1.$

Hence the equation with roots α and β is

$$x^2 - 2(\alpha + \beta)x + \alpha\beta = 0$$

$$\Rightarrow \quad x^2 - 2\cos n\theta\, x + 1 = 0$$

Example 16:

If α, β be the roots of $x^2 - 2x + 2 = 0$, prove that $\alpha^n + \beta^n = 2^{\frac{n}{2}+1}\cos\left(\frac{n\pi}{4}\right)$ and hence evaluate $\alpha^6 + \beta^6$.

Solution:

Putting n = 6, we get:

$$\alpha^6 + \beta^6 = 2^4\cos\left(\frac{6\pi}{4}\right) = 16\cos\left(\frac{3\pi}{2}\right) = 0.$$

Example 17:

If α, β be the roots of $x^2 - 2x + 4 = 0$, prove that $\alpha^n + \beta^n = 2^{n+1}\cos\frac{n\pi}{3}$ and hence evaluate $\alpha^6 + \beta^6$.

Solution:

Solving $x^2 - 2x + 4 = 0$, we get

$$x = \frac{2 \pm \sqrt{4-16}}{2} = 1 \pm i\sqrt{3}.$$

Let $1 = r\cos\theta$ and $\sqrt{3} = r\sin\theta$. Then $r = 2$, $\tan\theta = \sqrt{3}$ or $\theta = \pi/3$.

We have $\alpha = r(\cos\theta + i\sin\theta)$, $\beta = r(\cos\theta - i\sin\theta)$.

$\therefore\ \alpha^n = r^n(\cos n\theta + i\sin n\theta)$, $\beta^n = r^n(\cos n\theta - i\sin n\theta)$

Hence $\alpha^n + \beta^n = 2r^n\cos n\theta = 2.2^n\cos\frac{n\pi}{3} = 2^{n+1}\cos\frac{n\pi}{3}$.

Putting $n = 6$, we get $\alpha^6 + \beta^6 = 2^7\cos 2\pi = 128$.

Example 18:

If $\cos 2\theta + i\sin 2\theta = p$, $\cos 2\phi + i\sin 2\phi = q$, prove that

$$2\cos(\theta - \phi) = \sqrt{\frac{p}{q}} + \sqrt{\frac{q}{p}},\quad 2i\sin(\theta - \phi) = \sqrt{\frac{p}{q}} - \sqrt{\frac{q}{p}}.$$

Solution:

We have $\sqrt{p} = (\cos 2\theta + i\sin 2\theta)^{1/2} = \cos\theta + i\sin\theta$.

Similarly, $\sqrt{q} = \cos\phi + i\sin\phi$.

$$\therefore \quad \sqrt{\frac{p}{q}} = \frac{\cos\theta + i\sin\theta}{\cos\phi + i\sin\phi} = \cos(\theta - \phi) + i\sin(\theta - \phi),$$

and $\sqrt{\frac{q}{p}} = \cos(\phi - \theta) + i\sin(\phi - \theta) = \cos(\theta - \phi) - i\sin(\theta - \phi)$.

Hence $\sqrt{\frac{p}{q}} + \sqrt{\frac{q}{p}} = 2\cos(\theta - \phi)$, $\sqrt{\frac{p}{q}} - \sqrt{\frac{q}{p}} = 2i\sin(\theta - \phi)$.

Example 19:

If $(1 + x)^n = c_0 + c_1x + \ldots + c_nx^n$, n being a positive integer, prove that:

(i) $c_0 - c_2 + c_4 - c_6 + \dots = 2^{n/2} \cos n\pi/4,$

(ii) $c_1 - c_3 + c_5 - \dots = 2^{n/2} \sin n\pi/4,$

(iii) $c_0 + c_4 + c_8 + \dots = 2^{n-2} + 2^{(n-2)/2} \cos n\pi/4.$

Solution:

We have

$$(1 + x)^n = c_0 + c_1x + c_2x^2 + c_3x^3 + c_4x^4 + \dots$$

Putting n = 1, – 1, i, – i successively, we get

$$2^n = c_0 + c_1 + c_2 + c_3 + c_4 + \dots$$

$$0 = c_0 - c_1 + c_2 - c_3 + c_4 - \dots$$

$$(1 + i)^n = c_0 + c_1i - c_2 - c_3i + c_4 \dots$$

$$(1 - i)^n = c_0 - c_1i - c_2 + c_3i + c_4 \dots$$

Adding the above relations, we get

$$2^n + (1 + i)^n + (1 - i)^n = 4\,(c_0 + c_4 + c_8 + \dots)$$

$$\Rightarrow\ 2^n + 2^{n/2}\left(\cos\frac{n\pi}{4} + i\sin\frac{n\pi}{4}\right) + 2^{n/2}\left(\cos\frac{n\pi}{4} - i\sin\frac{n\pi}{4}\right)$$

$$= 4\,(c_0 + c_4 + c_8 + \dots)$$

$$\Rightarrow\ 2^n + 2.2^{n/2}\cos\frac{n\pi}{4} = 2^2\,(c_0 + c_4 + c_8 + \dots)$$

Dividing by 2^2 on both sides, we get:

$$c_0 + c_4 + c_8 + \dots = 2^{n-2} + 2^{\frac{n-2}{2}}\cos\frac{n\pi}{4}.$$

Example 20:

If $(\cos\theta + i\sin\theta)(\cos 2\theta + i\sin 2\theta)\dots(\cos n\theta + i\sin n\theta) = 1$, *prove that*

$$\theta = \frac{4k\pi}{n\,(n+1)},\ \textit{where } k \textit{ is any integer.}$$

Solution:

We have

$$(\cos\theta + i\sin\theta)(\cos 2\theta + i\sin 2\theta)\dots(\cos n\theta + i\sin n\theta)$$

$$= (\theta + 2\theta + \dots + n\theta) + i\sin(\theta + 2\theta + \dots + n\theta)$$

$$= \cos(1 + 2 + \dots + n)\,\theta + i\sin(1 + 2 + \dots + n)\,\theta$$

$$= \cos \frac{n(n+1)\theta}{2} + i \sin \frac{n(n+1)\theta}{2}$$

$$= \cos 2k\pi + i \sin 2k\pi, \text{ where } 2k\pi = \frac{n(n+1)\theta}{2}$$

$$= 1, \text{ where } \theta = \frac{4k\pi}{n(n+1)}, \text{ k being any integer.}$$

Example 21:

Prove that:

$$\left(\frac{-1+i\sqrt{3}}{2}\right) + \left(\frac{-1-i\sqrt{3}}{2}\right)^n$$

has value 2 if n = 3k and – 1 if n = 3k ± 1, where k is an integer.

Solution:

Let $r \cos \theta = -\frac{1}{2}$ and $r \sin \theta = \frac{\sqrt{3}}{2}$. Then $r^2 = 1 \Rightarrow r = 1$, $\tan \theta = -\sqrt{3} \Rightarrow \theta = 2\pi/3$. We have

$$\left(\frac{-1+i\sqrt{3}}{2}\right)^n + \left(\frac{-1-i\sqrt{3}}{2}\right)^n = r^n (\cos\theta + i \sin\theta)^n + r^n (\cos\theta - i \sin\theta)^n$$

$$= \cos n\theta + i \sin n\theta + \cos n\theta - i \sin n\theta \qquad (\because r = 1)$$

$$= 2 \cos n\theta = 2 \cos \left(\frac{2n\pi}{3}\right)$$

$$= 2 \cos 2k\pi, \text{ if } n = 3k \text{ (k any integer)}$$

$$= 2.1 = 2, \text{ if } n = 3k \text{ (k any integer)}$$

If $n = 3k \pm 1$, then

$$2 \cos \frac{2n\pi}{3} = 2 \cos\left(2k\pi \pm \frac{2\pi}{3}\right) = 2 \cos \frac{2\pi}{3} = -2.\frac{1}{2} = -1.$$

Example 22:

From the identity:

$$\frac{1}{x-a} - \frac{1}{x-b} \equiv \frac{a-b}{(x-a)(x-b)}, \textit{ prove that}$$

(i) $\cos(\theta + \alpha)\sin(\theta - \beta) - \cos(\theta + \beta)\sin(\theta - \alpha) = \sin(\alpha - \beta)\cos 2\theta,$

(ii) $\sin(\theta + \alpha)\sin(\theta - \beta) - \sin(\theta + \beta)\sin(\theta - \alpha)$
$= \sin(\alpha - \beta)\sin 2\theta.$

Solution:

Let $x = \cos 2\theta + i\sin 2\theta$, $a = \cos 2\alpha + i\sin 2\alpha$,
$b = \cos 2\beta + i\sin 2\beta$.

$\therefore \quad x - a = (\cos 2\theta - \cos 2\alpha) + i(\sin 2\theta - \sin 2\alpha)$

$= -2\sin(\theta + \alpha)\sin(\theta - \alpha) + 2i\cos(\theta + \alpha)\sin(\theta - \alpha)$

$= 2i\sin(\theta - \alpha)\{\cos(\theta + \alpha) + i\sin(\theta + \alpha)\}.$

Similarly,

$x - b = 2i\sin(\theta - \beta)\{\cos(\theta + \beta) + i\sin(\theta + \beta)\},$

$a - b = 2i\sin(\alpha - \beta)\{\cos(\alpha + \beta) + i\sin(\alpha + \beta)\}.$

Putting these values in the given idenity, we get

$$\frac{1}{2i\sin(\theta-\alpha)}[\cos(\theta + \alpha) - i\sin(\theta + \alpha)]$$

$$-\frac{1}{2i\sin(\theta-\beta)}[\cos(\theta + \beta) - i\sin(\theta + \beta)]$$

$$= \frac{2i\sin(\alpha-\beta)\left[\cos(\alpha+\beta) + i\sin(\alpha+\beta)\right]}{-4\sin(\theta-\alpha)\sin(\theta-\beta)\left\{\cos(2\theta+\alpha+\beta) + i\sin(2\theta+\alpha+\beta)\right\}}$$

Multiplying both sides by 2i, we get:

$$= \frac{\cos(\theta+\alpha) - i\sin(\theta+\alpha)}{\sin(\theta-\alpha)} - \frac{\cos(\theta-\beta) - i\sin(\theta+\beta)}{\sin(\theta-\beta)}$$

$$= \frac{\sin(\alpha-\beta)\left\{\cos(\alpha+\beta-2\theta-\alpha-\beta) + i\sin(\alpha+\beta-2\theta-\alpha-\beta)\right\}}{\sin(\theta-\alpha)\sin(\theta-\beta)}$$

$$= \frac{\sin(\alpha-\beta)(\cos 2\theta - i\sin 2\theta)}{\sin(\theta-\alpha)\sin(\theta-\beta)}.$$

Comparing the real parts on both the sides, we get

$$\frac{\cos(\theta+\alpha)}{\sin(\theta-\alpha)} - \frac{\cos(\theta+\beta)}{\sin(\theta-\beta)} = \frac{\sin(\alpha-\beta)\cos 2\theta}{\sin(\theta-\alpha)\sin(\theta-\beta)}$$

$\Rightarrow \cos(\theta + \alpha)\sin(\theta - \beta) - \cos(\theta + \beta)\sin(\theta - \alpha) = \sin(\alpha - \beta)\sin 2\theta.$

Similarly on comparing the imaginary parts, we get part (ii).

Example 23:

State Summation of Series for rational index and use it to solve the equation $z^7 + z = 0$.

Solution:

$Z^7 + z = 0 \Rightarrow z\,(z^6 + 1) = 0 \Rightarrow z = 0 \Rightarrow z^6 + 1 = 0.$

The six roots of $z^6 + 1 = 0$ are

$$+\,i,\ \frac{1}{2}\left(\sqrt{3} \pm i\right), \frac{1}{2}\left(-\sqrt{3} \pm i\right).$$

Example 24:

Express cos 7θ, sin 7θ in powers of cos θ and sin θ, and hence obtain an expression for tan 7θ in powers of tan θ.

Solution:

We have

$\cos 7\theta = \cos^7\theta - 21 \cos^5 \theta \sin^2 \theta + 35 \cos^3 \theta \sin^4 \theta - 7 \cos \theta \sin^6 \theta,$

$\sin 7\theta = 7 \cos^6 \theta \sin \theta - 35 \cos^4 \theta \sin^3 \theta + 21 \cos^2 \theta \sin^5 \theta - \sin^7 \theta.$

Hence

$$\tan 7\theta = \frac{7 \cos^6 \theta \sin \theta - 35 \cos^4 \theta \sin^3 \theta + 21 \cos^2 \theta \sin^5 \theta - \sin^7 \theta}{\cos^7 \theta - 21 \cos^5 \theta \sin^2 \theta + 35 \cos^3 \theta \sin^4 \theta - 7 \cos \theta \sin^6 \theta}$$

On dividing numerator and denominator by $\cos^7 \theta$, we get

$$\tan 7\theta = \frac{7 \tan \theta - 35 \tan^3 \theta + 21 \tan^5 \theta - \tan^7 \theta}{1 - 21 \tan^2 \theta + 35 \tan^4 \theta - 7 \tan^6 \theta}.$$

Example 25:

Show that

$$2^{11} \cos^7 \theta \sin^5 \theta = \sin 12\theta + 2 \sin 10\theta - 4 \sin 8\theta - 10 \sin 6\theta + 5 \sin 4\theta + 20 \sin 2\theta.$$

Solution:

Let $z = \cos \theta + i \sin \theta$. Then $1/z = \cos \theta - i \sin \theta$, $z + 1/z = 2 \cos \theta$, $z - 1/z = 2\, i \sin \theta$.

$$\therefore\quad (2 \cos \theta)^7 (2i \sin \theta)^5 = \left(z + \frac{1}{z}\right)^7 \left(z - \frac{1}{z}\right)^5$$

$$= \left(z+\frac{1}{z}\right)^2 \left(z^2-\frac{1}{z^2}\right)^5$$

$$= (z^2 + z^{-2} + 2)(z^{10} - 5z^6 + 10z^2 - 10z^{-2} + 5z^{-6} - z^{-10})$$

$$= (z^{12} - z^{-12}) + 2(z^{10} - z^{-10}) - 4(z^8 - z^{-8}) - 10(z^6 - z^{-6}) + 5(z^4 - z^{-4}) + 20(z^2 - z^{-2})$$

$$\Rightarrow 2^{12}.i.\cos^7\theta \sin^5\theta = 2i \sin 12\theta + 2(2i \sin 10\theta) - 4(2i \sin 8\theta) - 10(2i \sin 6\theta) + 5(2i \sin 4\theta) + 20(2i \sin 2\theta).$$

Hence $2^{11} \cos^7\theta \sin^5\theta = \sin 12\theta + 2\sin 10\theta - 4\sin 8\theta - 10\sin 6\theta + 5\sin 4\theta + 20\sin 2\theta$,

Example 26:

Find the sum to infinity of the series:

$$\sin\alpha + \frac{1}{2}\sin 3\alpha + \frac{1.3}{2.4}\sin 5\alpha + \ldots, \ -\pi < \alpha < \pi.$$

Solution:

Let $S = \sin\alpha + \frac{1}{2}\sin 3\alpha + \frac{1.3}{2.4}\sin 5\alpha + \ldots$

and $C = \cos\alpha + \frac{1}{2}\cos 3\alpha + \frac{1.3}{2.4}\cos 5\alpha + \ldots$

$$\therefore \quad C + iS = e^{i\alpha} + \frac{1}{2}e^{3i\alpha} + \frac{1.3}{2.4}e^{5i\alpha} + \ldots$$

$$= e^{i\alpha}\left[1+\frac{1}{2}e^{2i\alpha}+\frac{1.3}{2.4}e^{4i\alpha}+\ldots\right]$$

$$= e^{i\alpha}(1 - e^{2i\alpha})^{-1/2}$$

$$= \frac{\cos\alpha + i\sin\alpha}{[1-(\cos 2\alpha + i\sin 2\alpha)]^{1/2}}$$

$$= \frac{\cos\alpha + i\sin\alpha}{(2\sin^2\alpha - 2i\sin\alpha\cos\alpha)^{1/2}}$$

$$= \frac{\cos\alpha + i\sin\alpha}{\sqrt{2\sin\alpha}\,(\sin\alpha - i\cos\alpha)^{1/2}}$$

$$= \frac{\cos\alpha + i\sin\alpha}{\sqrt{2\sin\alpha}\,\{\cos(\pi/2-\alpha) - i\sin(\pi/2-\alpha)\}^{1/2}}$$

$$= \frac{\cos\alpha + i\sin\alpha}{\sqrt{2\sin\alpha}\,\{\cos(\pi/4-\alpha/2) - i\sin(\pi/4-\alpha/2)\}}$$

by DeMoivre's theorem

$$= \frac{1}{\sqrt{2\sin\alpha}}\left\{\cos\left(\alpha+\frac{\pi}{4}-\frac{\alpha}{2}\right) + i\sin\left(\alpha+\frac{\pi}{4}-\frac{\alpha}{2}\right)\right\}$$

$$= \frac{1}{\sqrt{2\sin\alpha}}\left\{\cos\left(\frac{\pi}{4}+\frac{\alpha}{2}\right) + i\sin\left(\frac{\pi}{4}+\frac{\alpha}{2}\right)\right\}.$$

...(1)

Equating the imaginary parts on both sides, we get

$$S = \sin\left(\frac{\pi}{4}+\frac{\alpha}{2}\right)\ (2\sin a)^{-1/2}.$$

Example 27:

Find the sum to infinity of the series

$$C: 1 + \frac{1}{2}\cos 2\theta - \frac{1}{2.4}\cos 4\theta + \frac{1.3}{2.4.6}\cos 6\theta - \ldots$$

Solution:

Let $S = \frac{1}{2}\sin 2\theta - \frac{1}{2.4}\sin 4\theta + \frac{1.3}{2.4.6}\cos 6\theta - \ldots$

Then $\quad C + iS = 1 + \frac{1}{2}e^{2i\theta} - \frac{1}{2.4}e^{4i\theta} + \frac{1.3}{2.4.6}e^{6i\theta} - \ldots$

$= (1 + e^{2i\theta})^{1/2} = = (1 + \cos 2\theta + i\sin 2\theta)^{1/2}$

$= (2\cos^2\theta + 2i\sin\theta\cos\theta)^{1/2}$

$= \sqrt{2\cos\theta}\ (\cos\theta/2 + i\sin\theta/2)$

Hence $C = \sqrt{2\cos\theta}\ \cos\theta/2.$

Example 28:

Sun the series: $\sin\theta\sin\theta + \sin^2\theta\sin 2\theta + \ldots + \sin^n\theta\sin n\theta$.

Solution:

Let $S = \sin\theta\sin\theta + \sin^2\theta\sin 2\theta + \ldots + \sin^n\theta\sin n\theta,$

$C = \sin\theta\cos\theta + \sin^2\theta\cos 2\theta + \ldots + \sin^n\theta\cos n\theta.$

$\therefore \quad C + iS = z\sin\theta + z^2\sin^2\theta + \ldots + z^n\sin^n\theta,\ z = \cos\theta + i\sin\theta$

$$= z \sin\theta\,(1 + z\sin\theta + \ldots + z^{n-1}\sin^{n-1}\theta)$$

$$= \frac{z\sin\theta\left(1 - z^n \sin^n\theta\right)}{1 - z\cos\theta}$$

$$= \frac{\sin\theta\left(\cos\theta + i\sin\theta\right)\left[1 - \sin^n\theta\left(\cos n\theta + i\sin n\theta\right)\right]}{1 - \cos\theta\left(\cos\theta + i\sin\theta\right)}$$

$$= \frac{\sin\theta\left(\cos\theta + i\sin\theta\right)\left[\left(1 - \sin^n\theta\cos n\theta\right) - i\sin^n\theta\sin n\theta\right]}{\sin^2\theta - i\sin\theta\cos\theta = -i\sin\theta\left(\cos\theta + i\sin\theta\right)}$$

$$= i\,[(1 - \sin^n\theta\cos n\theta) - i\sin^n\theta\sin n\theta]. \qquad \ldots(1)$$

Comparing the imaginary parts on both sides, we get

$$S = 1 - \sin^n\theta\cos n\theta.$$

Example 29:

Sum the series

3 sin α +5 sin 2α + 7 sin 3α + ... n terms.

Solution:

Let $S = 3\sin\alpha + 5\sin 2\alpha + 7\sin 3\alpha + \ldots + (2n+1)\sin n\alpha$

and $C = 1 + 3\cos\alpha + 5\cos 2\alpha + 7\cos 3\alpha + \ldots + (2n+1)\cos n\alpha$.

$\therefore \quad C + iS = 1 + 3e^{ia} + 5e^{2i\alpha} + 7e^{3i\alpha} + \ldots + (2n+1)\,e^{ni\alpha}$.

...(1)

Multiplying both sides by $e^{i\alpha}$, we get

$(C + iS)\,e^{i\alpha} = e^{i\alpha} + 3e^{2i\alpha} + 5e^{4i\alpha} + \ldots + (2n+1)\,e^{(n+1)i\alpha}$.

...(2)

Subtracting (2) from (1), we get

$$(C + iS)(1 - e^{i\alpha}) = 1 + 2\,(e^{i\alpha} + e^{2i\alpha} + \ldots + e^{ni\alpha}) - (2n+1)\,e^{(n+1)i\alpha}$$

$$= 1 + 2e^{i\alpha}\,\{1 + e^{i\alpha} + e^{2i\alpha} + \ldots e^{(n-1)\,i\alpha}\} - (2n+1)\,e^{(n+1)i\alpha}$$

$$= 1 + 2e^{i\alpha}\left[\frac{1 - e^{ni\alpha}}{1 - e^{i\alpha}}\right] - (2n+1)\,e^{(n+1)}\,i\alpha, \text{ if } e^{i\alpha} \neq 1$$

$$= \frac{1 + e^{i\alpha} - 2e^{(n+1)i\alpha} - (2n+1)\,e^{(n+1)i\alpha} + (2n+1)\,e^{(n+2)\,i\alpha}}{1 - e^{i\alpha}}$$

$$\therefore \quad C + iS = \frac{1 + e^{i\alpha} - (2n+3)\,e^{(n+1)i\alpha} + (2n+1)\,e^{(n+1)i\alpha}}{\left(1 - e^{i\alpha}\right)^2} \times \frac{e^{-i\alpha}}{e^{-i\alpha}}$$

$$= \frac{e^{-i\alpha} + 1 - (2n+3)\, e^{ni\alpha} + (2n+1)\, e^{(n+1)i\alpha}}{\{e^{-i\alpha/2}\,(1-e^{i\alpha})\}^2 = (e^{-i\alpha/2} - e^{i\alpha/2})^2}$$

$$= \frac{\cos\alpha - i\sin\alpha + 1 - (2n+3)\,(\cos n\alpha + i\sin n\alpha) + (2n+1)\,\{\cos(n+1)\,\alpha + i\sin(n+1)\,\alpha\}}{(2i\sin\alpha/2)^2 = -4\sin^2\alpha/2 = -2\,(1-\cos\alpha)}$$

Comparing imaginary parts on both sides, we get

$$S = \frac{\sin\alpha + (2n+3)\sin\alpha - (2n+1)\sin(n+1)\,\alpha}{2\,(1-\cos\alpha)}$$

if $\alpha \neq 2k\pi$ (k being any integer).

[$\because$ $e^{i\alpha} = 1 \Rightarrow \cos\alpha + i\sin\alpha = 1 \Rightarrow \cos\alpha = 1$ and $\sin\alpha = 0 \Rightarrow \alpha = 2k\pi$, k being any integer]

If $\alpha = 2k\pi$, the sum of the given series is obviously zero.

Example 30:

Sum the series to n terms:

(i) $\sin\alpha + 3\sin 2\alpha + 5\sin 3\alpha + \ldots$

(ii) $\cos\alpha + 3\cos 2\alpha + 5\cos 3\alpha + \ldots$

Solution:

Let $S = \sin\alpha + 3\sin 2\alpha + 5\sin 3\alpha + \ldots + (2n-1)\sin n\alpha$

and $C = \cos\alpha + 3\cos 2\alpha + 5\cos 3\alpha + \ldots + (2n-1)\cos n\alpha$.

$\therefore\ C + iS = e^{i\alpha} + 3e^{2i\alpha} + 5e^{3i\alpha} + \ldots + (2n-1)\,e^{ni\alpha}$. ...(1)

Multiplying both sides by $e^{i\alpha}$, we get:

$(C + iS)\,e^{i\alpha} = e^{2i\alpha} + 3e^{3i\alpha} + 5e^{4i\alpha} + \ldots + (2n-1)\,e^{(n+1)\,i\alpha}$.

...(2)

Subtracting (2) from (1), we get

$(C + iS)\,(1 - e^{i\alpha}) = e^{i\alpha} + 2e^{3i\alpha} + 2e^{3i\alpha} + \ldots + 2e^{ni\alpha} - (2n-1)\,e^{(n+1)i\alpha}$

$= e^{i\alpha} + 2e^{2i\alpha}\,\{1 + e^{i\alpha} + e^{2i\alpha} + \ldots + e^{(n-2)\,i\alpha}\} - (2n-1)\,e^{(n+1)\,i\alpha}$

$= e^{i\alpha} + 2e^{i\alpha}\left[\dfrac{1 - e^{(n-1)\,i\alpha}}{1 - e^{i\alpha}}\right] - (2n-1)\,e^{(n+1)\,i\alpha}$, if $e^{i\alpha} \neq 1$

$$= \frac{e^{i\alpha} + e^{2i\alpha} - 2e^{(n+1)\,i\alpha} - (2n-1)\,e^{(n+1)\,i\alpha} + (2n-1)\,e^{(n+2)\,i\alpha}}{(1 - e^{i\alpha})}$$

We have

$$C + iS = \frac{e^{i\alpha} + e^{2i\alpha} - (2n+1)e^{(n+1)i\alpha} + (2n-1)e^{(n+2)i\alpha}}{(1-e^{i\alpha})^2} \times \frac{e^{-i\alpha}}{e^{-i\alpha}}$$

$$= \frac{1 + e^{i\alpha} - (2n+1)e^{ni\alpha} + (2n-1)e^{(n+1)i\alpha}}{(e^{-i\alpha/2} - e^{i\alpha/2})^2 = (2i \sin\alpha/2)^2 = -4\sin^2\alpha/2}$$

$$= \frac{1 + (\cos\alpha + i\sin\alpha) - (2n+1)(\cos n\alpha + i\sin n\alpha) + (2n-1)\{\cos(n+1)\alpha + i\sin(n+1)\alpha\}}{-2(1-\cos\alpha)} \quad \ldots(1)$$

Comparing the imaginary parts on both the sides, we get

$$S = \frac{(2n+1)\sin n\alpha - (2n-1)\sin(n+1)\alpha - \sin\alpha}{2(1-\cos\alpha)},$$

provided $e^{i\alpha} \neq 1$ *i.e.*, $\alpha \neq 2k\pi$, k being any integer.

[$\because$ $e^{i\alpha} = 1 \Rightarrow \cos\alpha + i\sin\alpha = 1 \Rightarrow \cos\alpha = 1$ and $\sin\alpha = 0 \Rightarrow \alpha = 2k\pi$, k being any integer.]

If $\alpha = 2k\pi$, the sum of the series is obviously zero.

(ii) Comparing the real parts on both the sides of (1), we get

$$C = \frac{(2n+1)\cos n\alpha - (2n-1)\cos(n+1)\alpha - \cos\alpha - 1}{2(1-\cos\alpha)}, \quad a \neq 2k\pi$$

If $\alpha = 2k\pi$, the sum of the given series is

$1 + 3 + 5 + \ldots + (2n-1) = n^2$.

Example 31:

Sum the the series to n terms:

$\cos\alpha + 2\cos(\alpha+\beta) + 3\cos(\alpha+2\beta) + \ldots$

Solution:

Let

$$C = \cos\alpha + 2\cos(\alpha+\beta) + 3\cos(\alpha+2\beta) + \ldots + n\cos\left(a + \overline{n-1}\,\beta\right)$$

$$S = \sin\alpha + 2\sin(\alpha+\beta) + 3\sin(\alpha+2\beta) + \dots + n\sin\left(a+\overline{n-1}\,\beta\right)$$

$$\therefore\ C + iS = e^{i\alpha} + 2e^{i(\alpha+\beta)} + 3e^{i(\alpha+2\beta)} + \dots + ne^{i(a+\overline{n-1}\,\beta)}. \qquad \dots(1)$$

Multiplying (2) from (1), we get

$$(C + iS)(1 - e^{i\beta}) = e^{i\alpha} + e^{i(\alpha+\beta)} + e^{i(\alpha+3\beta)} + \dots + ne^{i(\alpha+n\beta)}$$

(2)

Subtracting (2) from (1), we get

$$(C + iS)(1 - e^{i\beta}) = e^{i\alpha} + e^{i(\alpha+\beta)} + e^{i(\alpha+2\beta)} + \dots e^{i(\varepsilon+\overline{n-1}\,\beta)} - ne^{i(\alpha+n\beta)}$$

$$= e^{i\alpha}\left[1 + e^{i\beta} + e^{2i\beta} + \dots + e^{(n-1)i\beta}\right] - ne^{i(\alpha+n\beta)}$$

$$= e^{i\alpha}\left(\frac{1-e^{ni\beta}}{1-e^{i\beta}}\right) - ne^{i(a+nb)}, \text{ provided } e^{i\beta} \neq 1$$

$$= \frac{e^{i\alpha} - (n+1)\,e^{i(\alpha+n\beta)} + ne^{i(\alpha+\overline{n+1}\,\beta)}}{1-e^{i\beta}}$$

We have

$$C + iS = \frac{e^{i\alpha} - (n+1)\,e^{i(\alpha+n\beta)} + e^{i(\alpha+\overline{n+1}\,\beta)}}{\left(1-e^{i\beta}\right)^2} \times \frac{\alpha^{-i\beta}}{e^{-i\beta}}$$

$$= \frac{e^{i(\alpha-\beta)} - (n+1)\,e^{i\left(\alpha+\overline{n-1}\,\beta\right)} + ne^{i(\alpha+n\beta)}}{\left(e^{-i\beta/2} - e^{i\beta/2}\right)^2 = \left(2i\sin\beta/2\right)^2 = -4\sin^2\beta/2}$$

Comparing the real parts on both the sides, we get

$$C = \frac{1}{4}\left[(n+1)\{\cos\alpha + (n-1)\beta\} - n\cos n(\alpha+n\beta) - \cos(\alpha-\beta)\right] \times \operatorname{cosec}^2\beta/2,$$

if $e^{i\beta} \neq 1$ *i.e.*, $\beta \neq 2k\pi$ (k being any integer).

If $\beta = 2k\pi$, the sum of the given series is

$$\cos\alpha + 2\cos\alpha + 3\cos\alpha + \dots + n\cos\alpha = \frac{n(n+1)}{2}\cos\alpha.$$

A Useful Formula

$$[x - (\cos\alpha + i\sin\alpha)][x - (\cos\alpha - i\sin\alpha)] = x^2 - 2x\cos\alpha + 1.$$

To verify this, we see that

$$\text{L.H.S.} = [(x - \cos\alpha) - i\sin\alpha][(x - \cos\alpha) + i\sin\alpha]$$

$$= (x - \cos\alpha)^2 - (i\sin\alpha)^2 = (x - \cos\alpha)^2 + \sin^2\alpha$$

$= x^2 - 2x \cos a + 1 = \text{R.H.S.}$

Example 32:

Solve the equation $z^8 + 1 = 0$ and deduce that

$$\cos 4\theta = 8 \prod_{p=0}^{3} \left[\cos \theta - \cos (2p+1) \frac{\pi}{8} \right]$$

Solution:

$z^8 + 1 = 0 \Rightarrow z = (-1)^{1/8} = (\cos \pi + i \sin \pi)^{1/8}$

$$\Rightarrow \quad z = \cos (2p + 1) \frac{\pi}{8} + i \sin (2p + 1) \frac{\pi}{8}, \; p = 0, 1..., 7.$$

$$\Rightarrow \quad z = \cos (2p + 1) \frac{\pi}{8} \pm i \sin (2p + 1) \frac{\pi}{8}, \; p = 0, 1, 2, 3.$$

$$\therefore \quad z^8 + 1 = \prod_{p=0}^{3} \left[z - \left(\cos (2p+1) \frac{\pi}{8} - i \sin (2p+1) \frac{\pi}{8} \right) \right]$$

$$\times \left[z - \left(\cos (2\pi+1) \frac{\pi}{8} - i \sin (2p+1) \frac{\pi}{8} \right) \right].$$

By the above formula, we see that

$$z^8 + 1 = \prod_{p=0}^{3} \left\{ z^2 - 2z \cos (2p + 1) \frac{\pi}{8} + 1 \right\}$$

Dividing both sides of (1) by z^4, we get

$$z^4 + z^{-4} = \prod_{p=0}^{3} \left\{ z - 2 \cos (2p+1) \frac{\pi}{8} + z^{-1} \right\}. \qquad ...(2)$$

On putting $z = \cos \theta + i \sin \theta$ in (2), we get

$$2 \cos 4\theta = \prod_{p=0}^{3} \left[2 \cos \theta - 2 \cos (2p+1) \frac{\pi}{8} \right] \quad \left(\begin{array}{l} \therefore \; z + z^{-1} = 2 \cos \theta \\ z^4 + z^{-4} = 2 \cos 4\theta \end{array} \right)$$

$$\text{Hence } \cos 4\theta = 8 \prod_{p=0}^{3} \left[\cos \theta - \cos (2p+1) \frac{\pi}{8} \right].$$

Example 33:

Solve the equation $z^{14} - 1 = 0$ and show that:

$$z^{14} - 1 = (z^2 - 1) \prod_{p=1}^{3} \left(z^2 - 2z \cos \frac{p\pi}{7} + 1 \right).$$

Hence show that: $\sin 7\theta = 64 \sin \theta \prod_{p=1}^{3}\left(\sin^2 \frac{p\pi}{7} - \sin^2\theta\right)$

Solution:

$z^{14} - 1 \Rightarrow z = (1)^{1/14} = (\cos 0 + i \sin 0)^{1/14}$

$$\Rightarrow \quad z = \cos \frac{2k\pi}{14} + i \sin \frac{2k\pi}{14}; \; k = 0, 1, \ldots 13$$

$$\therefore \quad z = \pm 1, \cos \frac{p\pi}{7} \pm i \sin \frac{p\pi}{7}, \; p = 1, 2, \ldots, 6.$$

Now $\quad z^{14} \Rightarrow 1 = (z^2 - 1) \prod_{p=1}^{6}\left(z^2 - 2z \cos p\pi/7 + 1\right).$

...(1)

Dividing both sides of (1) by z^7, we get

$$(z^7 - z^{-7}) = (z - z^{-1}) \prod_{p=1}^{6}\left(z - 2 \cos p\pi/7 + z^{-1}\right). \qquad \ldots(2)$$

On putting $z = \cos\theta + i \sin\theta$ in (2), we get

$$2i \sin 7\theta = 2i \sin\theta \prod_{p=1}^{6}\left(2\cos\theta - 2 \cos p\pi/7\right) \quad \left(\because \begin{array}{l} z + z^{-1} = 2 \cos\theta \\ z - z^{-1} = 2i \sin\theta \end{array}\right)$$

$$\Rightarrow \quad \sin 7\theta = 2^6 \sin\theta \prod_{p=1}^{3} (\cos\theta - \cos p\,\pi/7)(\cos\theta + \cos p\,\pi/7)$$

$$= 2^6 \sin\theta \prod_{p=1}^{3} (\cos^2\theta - \cos^2 p\pi/7)$$

$$= 2^6 \sin\theta \prod_{p=1}^{3} [(1 - \sin^2\theta) - (1 - \sin^2 p\pi/7)]$$

Hence $\sin 7\theta = 64 \sin\theta \prod_{p=1}^{3} (\sin^2 p\pi/7 = \sin^2\theta).$

Example 34:

Solve the equation $z^{10} - 1 = 0$ *and deduce that* $\sin 5\theta = 5 \sin\theta$

$$\left[1 - \frac{\sin^2\theta}{\sin^2(\pi/5)}\right]\left[1 - \frac{\sin^2\theta}{\sin^2(2\pi/5)}\right].$$

Solution:

$z^{10} - 1 = 0 \Rightarrow z = (\cos 0 + i \sin 0)^{1/10}$

$\therefore \quad z = \cos \dfrac{2p\pi}{10} + i \sin \dfrac{2\pi p}{10}$; i = 0, 1,....,9

$\Rightarrow \quad z = \cos \dfrac{p\pi}{5} + i \sin \dfrac{p\pi}{5}$; i = 0, 1,...,9

$= \pm 1, \cos p\pi/5 + i \sin p\pi/5$; p = 1, 2, 3, 4, 5, 6,7, 8, 9

$= \pm 1, \cos p\, \pi/5 \pm i \sin p\pi/5$; p = 1, 2, 3, 4.

Now $z^{10} - 1 = (z^2 - 1) \prod_{p=1}^{4} (z^2 - 2z \cos p\pi/5 + 1)$. ...(1)

Dividing both sides of (2) by z^5, we get

$(z^5 - z^{-5}) = (z - z^{-1}) \prod_{p=1}^{4} (z - 2 \cos p\, \pi/5 + z^{-1})$. ...(2)

Putting $z = \cos\theta + i \sin\theta$ in (2), we get

$$2i \sin 5\theta = 2i \sin\theta \prod_{p=1}^{4} (2\cos\theta - 2\cos p\pi/5)$$

$$\Rightarrow \quad \sin 5\theta = 2^4 \sin\theta \prod_{p=1}^{2} (\cos\theta - \cos p\pi/5)(\cos\theta + \cos p\pi/5)$$

$$= 2^4 \sin\theta \prod_{p=1}^{2} (\cos^2\theta - \cos^2 p\, \pi/5)$$

$$= 2^4 \sin\theta \prod_{p=1}^{2} (\sin^2 p\pi/5 - \sin^2\theta)$$

$$= 2^4 \sin\theta (\sin^2 \pi/5 - \sin^2\theta)(\sin^2 2\pi/5 - \sin^2\theta)$$

$$= 2^4 \sin^2 \pi/5 . \sin^2 2\pi/5 . \sin\theta \left[1 - \frac{\sin^2\theta}{\sin^2(\pi/5)}\right]\left[1 - \frac{\sin^2\theta}{\sin^2(2\pi/5)}\right]$$

$$= 5 \sin\theta \left[1 - \frac{\sin^2\theta}{\sin^2(\pi/5)}\right]\left[1 - \frac{\sin^2\theta}{\sin^2(2\pi/5)}\right]$$

$$\left[\because \sin\frac{\pi}{5} = \frac{\sqrt{10 - 2\sqrt{5}}}{4}, \sin\frac{2\pi}{5} = \frac{\sqrt{10 + 2\sqrt{5}}}{4}\right]$$

Example 35:

Prov that cos (2π/7), cos (4π/7), cos (6π/7) are the roots of the equation $8x^3 + 4x^2 - 4x - 1 = 0$.

Solution:

$\Rightarrow z^7 = 1 \Rightarrow z = (1)^{1/7} = (\cos 0 + i \sin 0)^{1/7}$

$$z = \cos \frac{2k\pi}{7} + i \sin \frac{2k\pi}{7} k = 0, 1, 2, \ldots, 6$$

$$\therefore \quad z = 1, \cos \frac{2k\pi}{7} \pm i \sin \frac{2k\pi}{7}; k = 1, 2, 3.$$

Using the formula given earlier, wre have

$$z^7 - 1 = (z - 1)[z^2 - 2z \cos (2\pi/7) + 1][z^2 - 2z \cos (4\pi/7) + 1] \times [z^2 - 2z \cos (6\pi/7) + 1].$$

Dividing both sides by z – 1, we obtain

$$z^6 + z^5 + z^4 + z^3 + z^2 + z + 1 = (z^2 - 2z \cos 2\pi/7 + 1) \times (z^2 - 2z \cos 4\pi/7 + 1)(z^2 - 2z \cos 6\pi/7 + 1).$$

Dividing the L.H.S. by z^3 and each of the three factors on the R.H.S. by z, we get

$$(z^3 + z^{-3}) + (z^2 + z^{-2}) + (z + z^{-1}) + 1$$
$$= (z + z^{-1} - 2 \cos 2\pi/7)(z + z^{-1} - 2 \cos 4\pi/7)(z + z^{-1} - 2 \cos 6\pi/7). \quad \ldots(1)$$

Let $z + z^{-1} = 2x$. Then $4x^2 = (z + z^{-1})^2 = z^2 + z^{-2} + 2$

$\Rightarrow z^2 + z^{-2} = 4x^2 - 2.$

Now $8x^3 = (z + z^{-1})^3 = z^3 + z^{-3} + 3zz^{-1}(z + z^{-1}) = z^3 + z^{-3} + 6x$

$\Rightarrow z^3 + z^{-3} = 8x^3 - 6x$. Then equation (1) now becomes

$$(8x^3 - 6x) + (4x^2 - 2) + 2x + 1$$
$$= (2x - 2 \cos 2\pi/7)(2x - 2\cos 4\pi/7)(2x - 2\cos 6\pi/7)$$

$$\Rightarrow 8x^3 + 4x^2 - 4x - 1 = 8(x - \cos 2\pi/7)(x - \cos 4\pi/7)(x - \cos 6\pi/7). \quad \ldots(2)$$

It follows that cos 2π/7, cos 4π/7, cos 6π/7 are the roots of the equation

$$8x^3 + 4x^2 - 4x - 1 = 0.$$

Example 36:

Solve the equation $x^7 - 1 = 0$, and deduce that

(i) $\cos\frac{2\pi}{7}+\cos\frac{4\pi}{7}+\cos\frac{8\pi}{7}=-\frac{1}{2}$,

(ii) $\cos\frac{2\pi}{7}\cos\frac{4\pi}{7}\cos\frac{8\pi}{7}=\frac{1}{8}$.

Solution:

From equation (2) and keeping in mind the relations between the roots coefficients, it follows that

$$\cos\frac{2\pi}{7}+\cos\frac{4\pi}{7}+\cos\frac{6\pi}{7}=-\frac{4}{8}=-\frac{1}{2},$$

$$\cos\frac{2\pi}{7}\cos\frac{4\pi}{7}\cos\frac{6\pi}{7}=\frac{1}{8}.$$

Example 37:

Solve the equation $z^7 + 1 = 0$, and deduce that

$$\cos\frac{\pi}{7}\cos\frac{3\pi}{7}\cos\frac{5\pi}{7}=-\frac{1}{8}.$$

Solution:

$z^7 + 1 = 0 \Rightarrow z = (-1)^{1/7} = (\cos\pi + i\sin\pi)^{1/7}$

$\Rightarrow \quad z = \cos(2k+1)\frac{\pi}{7} + i\sin(2k+1)\frac{\pi}{7};\ k = 0, 1, 2, \ldots, 6$

$\therefore \quad z = -1, \cos\frac{p\pi}{7} \pm i\sin\frac{p\pi}{7},\ p = 1, 3, 5$

(For k = 3, we obtain z = – 1)

We can write

$z^7 + 1 = (z+1)(z^2 - 2z\cos\pi/7 + 1)(z^2 - 2z\cos 3\pi/7 + 1)$
$\times (z^2 - 2z\cos 5\pi/7 + 1)$. ...(1)

Dividing both sides by z + 1, we obtain

$z^6 - z^5 + z^4 - z^3 + z^2 - z + 1 = (z^2 - 2z\cos\pi/7 + 1)$
$(z^2 - 2z\cos 3\pi/7 + 1) \times (z^2 - 2z\cos 5\pi/7 + 1)$.

Dividing the L.H.S. by z^3 and each of the three factors on the R.H.S. by z, we get

$(z^3 + z^{-3}) - (z^2 + z^{-2}) + (z + z^{-1}) - 1$

$= (z + z^{-1} - 2\cos\pi/7)(z + z^{-1} - 2\cos 3\pi/7)(z + z^{-1} - 2\cos 5\pi/7).$

Putting $z + z^{-1} = 2x$, we obtain

$(8x^3 - 6x) - (4x^2 - 2) + 2x - 1$

$= 8(x - \cos\pi/7)(x - \cos 3\pi/7)(x - \cos 5\pi/7)$

Thus $\cos\pi/7$, $\cos 3\pi/7$, $\cos 5\pi/7$ are the roots of the equation

$$8x^3 - 4x^2 - 4x + 1 = 0.$$

Further $\cos\frac{\pi}{7} + \cos\frac{3\pi}{7} + \cos\frac{5\pi}{7} = \frac{4}{8} = \frac{1}{2}$,

and $\cos\frac{\pi}{7}\cos\frac{3\pi}{7}\cos\frac{5\pi}{7} = -\frac{1}{8}$.

Example 38:

Solve the equation $z^9 + 1 = 0$ and hence obtain the equation whose roots are

$$\cos\frac{\pi}{9}, \cos\frac{3\pi}{9}, \cos\frac{5\pi}{7}, \cos\frac{7\pi}{9}.$$

Solution:

$z^9 + 1 = 0 \Rightarrow z = (\cos\pi + i\sin\pi)^{1/9}$

$\Rightarrow \quad z = \cos\frac{(2k+1)\pi}{9} + i\sin\frac{(2k+1)\pi}{9}$; $k = 0, 1, ..., 8$

$\Rightarrow \quad z = -1, \cos\frac{p\pi}{9} \pm i\sin\frac{p\pi}{9}$; $p = 1, 3, 5, 7.$

$z^9 + 1 = (z + 1)(z^2 - 2z\cos\pi/9 + 1)(z^2 - 2z\cos 3\pi/9 + 1)$
$\times (z^2 - 2z\cos 5\pi/9 + 1)(z^2 - 2z\cos 7\pi/9 + 1)$

Dividing both sides by $z + 1$, we get

$z^8 - z^7 + z^6 - z^5 + z^4 - z^3 + z^2 - z + 1$

$= (z^2 - 2z\cos\pi/9 + 1)(z^2 - 2z\cos 3\pi/9 + 1)$
$\times (z^2 - 2z\cos 5\pi/9 + 1)(z^2 - 2z\cos 7\pi/9 + 1).$

Dividing the L.H.S. by z^4 and each of the four factors on the R.H.S. by z, we get

$(z^4 + z^{-4}) - (z^3 + z^{-3}) + (z^2 + z^{-2}) - (z + z^{-1}) + 1$

$= (z + z^{-1} - 2\cos\pi/9)(z + z^{-1} - 2\cos 3\pi/9)$
$(z + z^{-1} - 2\cos 5\pi/9) \times (z + z^{-1} - 2\cos 7\pi/9).$

Putting $z + z^{-1} = 2x$, we get

$$[16x^4 - 4(4x^2 - 2) - 6] - (8x^3 - 6x) + (4x^2 - 2) - 2x + 1$$
$$= 2^4 (x - \cos \pi/9)(x - \cos 3\pi/9)(x - \cos 5\pi/9)\, 9x - \cos 7\pi/9).$$

Hence $\cos \pi/9$, $\cos 3\pi/9$, $\cos 5\pi/9$, $\cos 7\pi/9$ are the roots of the equation

$$16x^4 - 16x^2 + 2 - 8x^3 + 6x + 4x^2 - 2x - 1 = 0$$

$$\Rightarrow \quad 16x^4 - 8x^3 - 12x^2 + 4x + 1 = 0.$$

Example 39:

Solve the equation $z^7 + 1 = 0$, and show that

$$(1 + y)^7 + (1 - y)^7 = 14\left(y^2 + \tan^2 \frac{\pi}{14}\right)\left(y^2 + \tan^2 \frac{3\pi}{14}\right)\left(y^2 + \tan^2 \frac{5\pi}{14}\right)$$

Deduce that: $\tan^2 \dfrac{\pi}{14} + \tan^2 \dfrac{3\pi}{14} + \tan^2 \dfrac{5\pi}{14} = 5$.

Solution:

The equation (2) of Example can be written as

$$z^7 + 1 = (z + 1) \prod_{p=0}^{2} \left[z^2 - 2z \cos (2p+1) \frac{\pi}{7} + 1\right]$$

Putting $z = \dfrac{1+y}{1-y} \Rightarrow z^7 + 1 = \dfrac{(1+y)^7 + (1-y)^7}{(1-y)^7}$,

$$z^2 + 1 = \frac{(1+y)^2 + (1-y)^2}{(1-y)^2} = \frac{2(1+y^2)}{(1-y)^2}, \quad z + 1 = \frac{2}{1-y}.$$

Using these relations in (2), we obtain

$$\frac{(1+y)^7 + (1-y)^7}{(1-y)^7} = \left(\frac{2}{1-y}\right) \prod_{p=0}^{2} \left[\frac{2(1+y^2)}{(1-y)^2} - \frac{2(1+y)}{(1-y)} \cos (2p+1) \frac{\pi}{7}\right]$$

$$= \left(\frac{2}{1-y}\right)\left[\frac{2}{(1-y)^2}\right]^3 \prod_{p=0}^{2} \left[(1+y^2) - (1-y^2) \cos (2p+1) \frac{\pi}{7}\right]$$

$$\Rightarrow \quad (1 + y)^7 + (1 - y)^7$$

$$= 2^4 \prod_{p=0}^{2} \left[\left\{1 - \cos (2p+1) \frac{\pi}{7}\right\} + y^2 \left\{1 + \cos(2p+1) \frac{\pi}{7}\right\}\right]$$

$$= 2^4 \prod_{p=0}^{2} \left[2 \sin^2 (2p+1) \frac{\pi}{14} + 2y^2 \cos^2 (2p+1) \frac{\pi}{14}\right]$$

$$= 2^7 \prod_{p=0}^{2} \left[\sin^2 (2p+1)\frac{\pi}{14} + y^2 \cos^2 (2p+1)\frac{\pi}{14}\right]. \qquad ...(3)$$

Dividing both sides of (2) by z + 1, we obtain

$$z^6 - z^5 + z^4 - z^3 + z^2 - z + 1 = \prod_{p=0}^{2} \left[z^2 - 2z\cos(2p+1)\frac{\pi}{7} + 1\right]$$

Putting z = – 1 in the above equation, we obtain

$$7 = \prod_{p=0}^{2} 2\left\{1+\cos(2p+1)\frac{\pi}{7}\right\} = \prod_{p=0}^{2} 2^2\cos^2(2p+1)\frac{\pi}{14}$$

$$\Rightarrow \; 7 = 2^6 \prod_{p=0}^{2} \cos^2 (2p+1)\frac{\pi}{14}. \qquad ...(4)$$

Dividing both sides of (3) by those of (4), we get

$$(1 + y)^7 + (1 - y)^7 = 14 \prod_{p=0}^{2} \left[y^2 + \tan^2 (2p+1)\frac{\pi}{14}\right]$$

$$= 14 \left(y^2 + \tan^2\frac{\pi}{14}\right)\left(y^2 + \tan^2\frac{3\pi}{14}\right)\left(y^2 + \tan^2\frac{5\pi}{14}\right).$$

Equating the coefficients of y^4 on both sides, we have $2\left(7_{c4}\right) = 14$

$$\left(\tan^2\frac{\pi}{14} + \tan^2\frac{3\pi}{14} + \tan^2\frac{5\pi}{14}\right)$$

Hence $\tan^2\frac{\pi}{14} + \tan^2\frac{3\pi}{14} + \tan^2\frac{5\pi}{14} = \frac{1}{7}\left(\frac{7!}{4!\,3!}\right) = 5.$

Example 40:

Solve the equation $(1 + w)^9 - (1 - w)^9 = 0$ and hence obtain the equation whose roots are

$$\tan^2\frac{\pi}{9},\ \tan^2\frac{2\pi}{9},\ \tan^2\frac{3\pi}{9},\ \tan^2\frac{4\pi}{9}$$

Deduce that: $\tan\frac{\pi}{9}\tan\frac{2\pi}{9}\tan\frac{3\pi}{9}\tan\frac{4\pi}{9} = 3.$

Solution:

Consider $z^9 - 1 = 0 \Rightarrow z = (\cos 0 + i \sin 0)^{1/9}$

$\Rightarrow \; z = \cos 2p\pi/9 + i \sin 2p\pi/9,\ p = 0, 1, ..., 8$

$\Rightarrow$ $z = 1, \cos 2p\pi/9 \pm i \sin 2p\pi/9, p = 1, 2, 3, 4.$

$$\therefore z^9 - 1 = (z-1)\prod_{p=1}^{4}(z^2 - 2z\cos 2p\pi/9 + 1).$$

Putting $z = \dfrac{1+w}{1-w}$, we obtain

$$\left(\frac{1+w}{1-w}\right)^9 - 1 = \left(\frac{1+w}{1-w} - 1\right)\prod_{p=1}^{4}\left[\left(\frac{1+w}{1-w}\right)^2 - 2\left(\frac{1+w}{1-w}\right)\cos\frac{2p\pi}{9} + 1\right]$$

$$\Rightarrow (1+w)^9 - (1-w)^9 = 2^5 w \prod_{p=1}^{4}\left[(1+w^2) - (1-w^2)\cos\frac{2p\pi}{9}\right]$$

$$= 2^5 w \prod_{p=1}^{4}\left[\left(1-\cos\frac{2p\pi}{9}\right) + w^2\left(1+\cos\frac{2p\pi}{9}\right)\right]$$

$$\therefore (1+w)^9 - (1-w)^9 = 2^9 w \prod_{p=1}^{4}\left[\sin^2\frac{p\pi}{9} + w^2\cos^2\frac{p\pi}{9}\right]. \quad ...(1)$$

Now $(1+w)^9 - (1-w)^9$

$$\left(1+9_{c1}w+9_{c2}w^2+9_{c3}w^3+9_{c4}w^4+9_{c5}w^5+9_{c6}w^6+9_{c7}w^7+9_{c8}w^8+w^9\right)$$

$$-\left(1-9_{c1}w+9_{c2}w^2-9_{c3}w^3+9_{c4}w^4-9_{c5}w^5+9_{c6}w^6-9_{c7}w^7+9_{c8}w^8-w^9\right)$$

$$= 2\left(9_{c1}w+9_{c3}w^3+9_{c5}w^5+9_{c7}w^7+w^9\right)$$

$$= 2(9w + 84w^3 + 126w^5 + 36w^7 + w^9). \quad ...(2)$$

Putting (2) in (1) and dividing by 2w, we get

$$9 + 84w^2 + 126w^4 + 36w^6 + w^8 = 2^8 \prod_{p=1}^{4}\left[\sin^2\frac{p\pi}{9} + w^2\cos^2\frac{p\pi}{9}\right]$$

$$= 2^8 \prod_{p=1}^{4}\cos^2\frac{p\pi}{9}\left[w^2 + \tan^2\frac{p\pi}{9}\right] \quad ...(3)$$

Putting $w^2 = -z$ in (3); we see that

$\tan^2(\pi/9)$, $\tan^2(2\pi/9)$, $\tan^2(3\pi/9)$, $\tan^2(4\pi/9)$ are the roots of

$$9 - 84z + 126z^2 - 36z^3 + z^4 = 0$$

$$\Rightarrow z^4 - 36z^3 + 126z^2 - 84z + 9 = 0. \quad ...(4)$$

(ii) From equation (4), it follows that

$$\tan^2\frac{\pi}{9}\tan^2\frac{2\pi}{9}\tan^2\frac{3\pi}{9}\tan^2\frac{4\pi}{9}=9.$$

Hence $\tan\frac{\pi}{9}\tan\frac{2\pi}{9}\tan\frac{3\pi}{9}\tan\frac{4\pi}{9}=3.$

Example 41:

Solve the equation $z^9 - 1 = 0$, and deduce that

(i) $\sin\frac{\pi}{9}\sin\frac{2\pi}{9}\sin\frac{3\pi}{9}\sin\frac{4\pi}{9}=\frac{3}{16}$,

(ii) $\tan\frac{\pi}{9}\tan\frac{2\pi}{9}\tan\frac{3\pi}{9}\tan\frac{4\pi}{9}=3.$

Solution:

We have seen that

$$\tan\frac{\pi}{9}\tan\frac{2\pi}{9}\tan\frac{3\pi}{9}\tan\frac{4\pi}{9}=3 \qquad ...(5)$$

Comparing the coefficients w^8 on both sides of (3), we get

$$1 = 2^8\cos^2\frac{\pi}{9}\cos^2\frac{2\pi}{9}\cos^2\frac{3\pi}{9}\cos^2\frac{4\pi}{9}$$

$$\Rightarrow \cos\frac{\pi}{9}\cos\frac{2\pi}{9}\cos\frac{3\pi}{9}\cos\frac{4\pi}{9}=\frac{1}{2^4}=\frac{1}{16}. \qquad ...(6)$$

Multiplying the respective sides of (5) and (6), we get

$$\sin\frac{\pi}{9}\sin\frac{2\pi}{9}\sin\frac{3\pi}{9}\sin\frac{4\pi}{9}=\frac{3}{16}.$$

Example 42:

Solve the equation $(1 + z)^{11} - (1 - z)^{11} = 0$, and hence obtain the equation whose roots are

$$\tan^2\frac{\pi}{11}, \tan^2\frac{2\pi}{11},\ldots, \tan^2\frac{5\pi}{11}.$$

Solution:

Consider $x^{11} - 1 = 0$

$$\Rightarrow x = \cos\frac{2p\pi}{11}+i\sin\frac{p\pi}{11}, \; p = 0, 1,\ldots, 10$$

$$\Rightarrow \quad x = 1,\ \cos\frac{2p\pi}{11} \pm i\sin\frac{2p\pi}{11};\ p = 1, 2, 3, 4, 5.$$

$$\therefore \quad x^{11} - 1 = (x-1)\prod_{p=1}^{5}\left(x^2 - 2x\cos\frac{2p\pi}{11} + 1\right).$$

Putting $x = \dfrac{1+z}{1-z}$ we obtain

$$\left(\frac{1+z}{1-z}\right)^{11} - 1 = \left(\frac{1+z}{1-z} - 1\right)\prod_{p=1}^{5}\left\{\left(\frac{1+z}{1-z}\right)^2 - 2\left(\frac{1+z}{1-z}\right)\cos\frac{2p\pi}{11} + 1\right\}$$

$$\Rightarrow \quad (1+z)^{11} - (1-z)^{11} = 2^6.z\prod_{p=1}^{5}\left\{(1+z^2) - (1-z^2)\cos\frac{2p\pi}{11}\right\}$$

$$\Rightarrow \quad (1+z)^{11} - (1-z)^{11} = 2^{11}.z\prod_{p=1}^{5}\left\{\sin^2\frac{p\pi}{11} + z^2\cos^2\frac{p\pi}{11}\right\}.$$

Expanding $(1+z)^{11}, (1-z)^{11}$ by Binomial theorem and dividing throughout by 2z, we obtain

$$11 + 165z^2 + 462z^4 + 330z^6 + 55z^8 + z^{10}$$

$$= 2^{10}\prod_{p=1}^{5}\left\{\sin^2\frac{p\pi}{11} + z^2\cos^2\frac{p\pi}{11}\right\}$$

$$= 2^{10}\prod_{p=1}^{5}\cos^2\frac{p\pi}{11}\left\{\tan^2\frac{p\pi}{11} + z^2\right\}.$$

Put $z^2 = -x$, we see that

$\tan^2\dfrac{\pi}{11}, \tan^2\dfrac{2\pi}{11}, \ldots, \tan^2\dfrac{5\pi}{11}$ are the roots of the equation

$$11 - 165x + 462x^2 - 330x^3 + 55x^4 - x^5 = 0$$

$$\Rightarrow \quad x^5 - 55x^4 + 330x^3 - 462x^2 + 165x - 11 = 0.$$

EXERCISES

1. $64\sin^7\theta = -\sin 7\theta + 7\sin 5\theta - 21\sin 3\theta + 35\sin\theta$.
2. $\cos 2\theta + \cos 4\theta + \cos 6\theta + \ldots$ **Ans.** $\sin n\theta\cos(n+1)\theta\ \text{cosec}\ \theta$
3. $\sin 2\theta + \sin 4\theta + \sin 6\theta + \ldots$ **Ans.** $\sin n\theta\sin(n+1)\theta\ \text{cosec}\ \theta$.
4. $\sin(\pi/2n) + \sin(3\pi/2n) + \sin(5\pi/2n) + \ldots$ **Ans.** $\text{cosec}\ (\pi/2n)$
5. $\sin\theta + \dfrac{1}{3}\sin 2\theta + \dfrac{1}{3^2}\sin 3\theta + \ldots$

Ans. $[\sin n\theta - 3 \sin (n + 1)\theta + 3^{n+1} \sin \theta] \div 3^{n-1}.2 (5 - 3 \cos \theta)$.

6. Show that

$\sin 6\theta = 6 \cos^5 \theta \sin \theta - 20 \cos^3 \theta \sin^3 \theta + 6 \cos \theta \sin^5 \theta$.

7. Express $\cos 8\theta$ in terms of powers of $\cos \theta$.

8. Show that $(\sin n\theta)/\sin \theta$

$$= (2 \cos \theta)^{n-1} - (n - 2)(2 \cos \theta)^{n-3} + \frac{(n-3)(n-4)}{1.2}(2 \cos \theta)^{n-5} + \ldots + \text{last term}$$

Giving reasons show that the last term is $(-1)^{(n-1)/2}$ if n is odd and $(-1)^{(n/2)-1} (n \cos \theta)$ if n is even

9. Show that $\sin \theta \cos^2 \theta = \frac{1}{4}\left[(1+3)\theta - (1+3^3)\frac{\theta^3}{3!} + (1+3^5)\frac{\theta^5}{5!} - \ldots\right]$.

10. Prove that $\cos^3 x = \frac{1}{4}\left[x - \frac{x^2}{2!}(3^2+3) + \frac{x^4}{4!}(3^4+3) - \ldots\right]$.

11. Show that

$$\sin^3\theta = \frac{3}{4}\left[\frac{(3^2-1)}{3!}\theta^3 - \frac{(3^4-1)}{5!}\theta^5 + \ldots + \frac{(-1)^{n-4}(3^{2n}-1)}{(2n+1)!}\theta^{2n-1} + \ldots\right]$$

12. If $\frac{\sin \theta}{\theta} = \frac{863}{864}$, show that $\theta = 4°47'$ nearly.

13. If $\frac{\sin \theta}{\theta} = \frac{1349}{1350}$, prove that the angle q is very nearly equal to $\frac{1}{15}$ radians.

14. Prove that:

$$\cos\frac{\pi}{11} + \cos\frac{3\pi}{11} + \cos\frac{5\pi}{11} + \cos\frac{7\pi}{11} + \cos\frac{9\pi}{11}, \cos\frac{11\pi}{13} = \frac{1}{2}$$

15. From the equation whose roots are

$$\cos\frac{\pi}{13}, \cos\frac{3\pi}{13}, \cos\frac{5\pi}{13}, \cos\frac{7\pi}{13}, \cos\frac{9\pi}{13}, \cos\frac{11\pi}{13}$$

16. Prove that $64 \sin^5 \theta \cos^2 \theta = \sin 7\theta - 3 \sin 5\theta + \sin 3\theta + 5 \sin \theta$

17. Show that

$64 \cos^7 \theta = \cos 7\theta + 7 \cos 5\theta + 21 \cos 3\theta + 35 \cos \theta$.

18. Prove that

$$\sin^6 \theta = -\frac{1}{32}[\cos 6\theta - 6 \cos 4\theta + 15 \cos 2\theta - 10].$$

19. Expand $\sin^5 \theta \cos^2 \theta$ in a series of sines of multiples of θ.
20. Express $\cos^3 \theta \sin^6 \theta$ as a series of cosines of multiples of θ.
21. Expand $\sin^3 \theta \cos^3 \theta$ in a series of sines of multiples of θ.
22. Expand $\cos^6 \theta \sin^3 \theta$ in a series of cosines of multipls of θ.
23. Expand $\sin^3 \theta \cos^8 \theta$ in a series of sines of muliples of θ.
24. Prove that $\cos^6 \theta \sin^2$

$$= -\frac{1}{128}[\cos 8\theta + 4\cos 6\theta + 4\cos 4\theta - 4\cos 2\theta - 5].$$

25. Express $\dfrac{(\cos\theta + i\sin\theta)^7}{(\cos 2\theta + i\sin 2\theta)^2}$ in the form $x + iy$.

[**Ans.** $\cos 3\theta + i\sin 3\theta$]

26. Express $\dfrac{\{\cos \pi/4 + i\sin \pi/4\}^5}{\{\cos \pi/3 + i\sin \pi/3\}^4}$ in the form $x + iy$.

[**Ans.** $\cos(\pi/12) - i\sin(\pi/12)$]

27. Simplify $\dfrac{(\cos 2\theta + i\sin 2\theta)^3 (\cos\theta - i\sin\theta)^{-4}}{(\cos 3\theta + i\sin 3\theta)^4 (\cos 4\theta + i\sin 4\theta)^{-2}}$

[**Ans.** $\cos 6\theta + i\sin 6\theta$]

28. If $2\cos\theta = \alpha + 1/a$, $2\cos\phi = \beta + 1/b$; show that

$$\frac{1}{2}\left(ab + \frac{1}{ab}\right) = \cos(\theta + \phi).$$

29. Prove tha $\{(\cos\alpha + i\sin\alpha) - (\cos\beta + i\sin\beta)\}^8 +$

$$\{(\cos\alpha - i\sin\alpha) - (\cos\beta - i\sin\beta)\}^8 = 2^8 \frac{1}{2}(\alpha - \beta)\cos 4(\alpha + \beta).$$

30. Prove Summation of Series when index is a negative integer.
31. $z^6 = 1$. Ans. $\pm 1, \pm\frac{1}{2} \pm i\left(\sqrt{3/2}\right)$.
32. $z^7 + 1 = 0$. [**Ans.** $\cos(p\pi/7) + i\sin(p\pi/7)$, p=1, 3, 5, 7, 9, 11, 13]
33. $z^8 + 1 = 0$. [**Ans.** $\cos(p\pi/8) + i\sin(p\pi/8)$, p=1, 3, 5, 7, 9, 11, 13, 15]
44. $z^4 + z^3 + z^2 + 1 = 0$.

    [**Hint:** L.H.S. $= (z^5 - 1)/(z - 1)$. Now solve $z^5 - 1 = 0$ and discard the value $z = 1$]. **Ans.** $\cos(p\pi/5) + i\sin(p\pi/5)$, p = 2, 4, 6, 8.

35. $x^8 - x^6 + x^4 - x^2 + 1 = 0$.

    [**Hint:** Multiply by $(x^2 + 1)$ to get $x^{10} + 1 = 0$ and reject the values given by $x^2 + 1 = 0$ *i.e.*, $\pm i$]

Ans. $\cos(p\pi/10) + i\sin(p\pi/10)$; $p = 1, 3, 7, 9, 11, 13, 17, 19$.

36. $x^7 + x^4 + x^3 + 1 = 0$.

[**Hint:** L.H.S. $= (x^4 + 1)(x^3 + 1)$.

Ans. $-1, \frac{1}{2} \pm i\left(\sqrt{3/2}\right); \frac{1}{\sqrt{2}}(\pm i \pm i)$.

37. $(z + 1)^5 + (z - 1)^5 = 0$. **Ans.** $0, \pm i\cot(p\theta/10)$; $p = 1, 3$.

38. $(4 + z)^5 - (4 - z)^5 = 0$ **Ans.** $0, \pm 4i\cot(\pi/10), \pm 4i\cot(3\pi/10)$.

39. If $x = \cos(2\pi/5) + i\sin(2\pi/5)$, show that $1 + x^n + x^{2n} + x^{3n} + x^{in}$ is 5 when n is multiple of 5 and 0 when n is not a multiple of 5.

40. Solve $x^m = 1$, m being a positive integer, and show that the roots are in G.P. Hence, show that the sum of qth powers of the roots is m or zero according as θ is a multiple of m or not.

41. $\cos 4\theta = \cos^4\theta - 6\cos^2\theta\sin^2\theta + \sin^4\theta$.

42. $\sin 4\theta = 4\cos^3\theta\sin\theta - 4\cos\theta\sin^3\theta$.

43. $\tan 4\theta = \dfrac{4\tan\theta - 4\tan^3\theta}{1 - 6\tan^2\theta + \tan^4\theta}$.

44. $8\sin^4\theta = \cos 4\theta - 4\cos 2\theta + 3$.

45. $8\cos^4\theta = \cos 4\theta + 4\cos 2\theta + 3$.

46. $2\cos 2\theta + 2^2\cos 4\theta + 2^3\cos 6\theta + \ldots$

[**Hint:** $C + iS = z + z^2 + z^3 + \ldots + z^n$, where $z = 2e^{2i\theta}$.

Ans. $2[\cos 2\theta - 2 - 2^n\cos 2(n + 1)\theta + 2^{n+1}\cos 2n\theta]$
$\div (5 - 4\cos 2\theta)$.

47. Sum the series: $\cos\alpha + \frac{1}{2}\cos 2\alpha + \frac{1}{2^2}\cos 3\alpha + \ldots +$ ad. inf.

[**Hint:** $C + iS = e^{ia} + \frac{1}{2}e^{2i\alpha} + \frac{1}{2^2}e^{3i\alpha} + \ldots +$ ad. inf.]

This is a geometric series with c.r. $= \frac{1}{2}e^{i\alpha}$

$$\therefore \quad C + iS = \frac{e^{i\alpha}}{1 - \frac{1}{2}e^{i\alpha}} = \frac{2(\cos\alpha + i\sin\alpha)}{2 - (\cos\alpha + i\sin\alpha)}$$

$$= \frac{(2\cos\alpha + 2i\sin\alpha)\left[(2 - \cos\alpha) + i\sin\alpha\right]}{(2 - \cos\alpha)^2 + \sin^2\alpha = 5 - 4\cos\alpha} \qquad \ldots(1)$$

$\therefore \quad C = [2\cos\alpha(2 - \cos\alpha) - 2\sin^2\alpha]/(5 - 4\cos\alpha)$

$= (4\cos\alpha - 2)/(5 - 4\cos\alpha)$.

48. Solve $x^6 + 1 = 0$.

49. Solve $x^7 - 1 = 0$.

50. Prove that the sum of the pth powers of the roots of the equation $x^n = 1$, where n is a positive integer is 0 when p is not a multiple of n and n when p is a multiple of n.

51. Find the seven seventh roots of unity and prove that these roots are in geometrical progression and the sum of their powers always vanishes unless n is a multiple of 7 (n being any integer).

52. Prove that, if x denotes any nth roots of unity, then $1 + x + x^2 + x^3 + \dots + x^n = 0$.

53. Discuss all the values of $(\cos\theta + i\sin\theta)^{1/\theta}$, where θ is positive integer.

54. If A and B are the roots $x^2 + 4 = 2x$, find the value of $A^n + B^n$.

55. Solve the equation $4x^4 = \sqrt{3+i}$.

Ans. $\frac{1}{2^{1/4}}\left[\cos\frac{r\pi}{24} + i\sin\frac{r\pi}{24}\right]$, where r = 1, 13, 25. 37

$$\left[\textbf{Hint. } x^4 = \frac{\sqrt{3+i}}{4} = \frac{1}{2}\left(\frac{\sqrt{3}}{2} + i.\frac{1}{2}\right) = \frac{1}{2}\left(\cos\frac{\pi}{6} + i\sin\frac{\pi}{6}\right)\right]$$